THE BOTANICAL GARDENS

AT THE HUNTINGTON

By

WALTER HOUK

Principal Photographer
DON NORMARK

Picture Commentaries
ROSEMARY VEREY

Contributors
JAMES P. FOLSOM, DIRECTOR
JOE CLEMENTS, SHIRLEY KERINS, CLAIR MARTIN,
NORAMELIA MARTINEZ, KATHY MUSIAL,
DAPHNE L. TRAGER, JOHN N. TRAGER,
JERROLD TURNEY

Supervising Editor
PEGGY PARK BERNAL

HUNTINGTON LIBRARY, SAN MARINO
HARRY N. ABRAMS, INC., NEW YORK

Copyright 1996 by the
Henry E. Huntington Library and Art Gallery
All rights reserved.

Printed in Singapore
ISBN 0-87328-155-5 (Huntington Library)
ISBN 0-8109-6316-7 (Abrams)

Library of Congress Cataloging-in-Publication data

Normark, Don, 1928–
    The Botanical Gardens at the Huntington / principal
photography by Don Normark ; contributors, James Folsom
... [et al.] ; supervising editor, Peggy Park Bernal.
        p.   cm.
    Includes bibliographical references (p.   ) and index.
    ISBN 0-87328-155-1 (Huntington Library : hardcover).
    — ISBN 0-8109-6316-7 (Abrams : hardcover)
    1. Huntington Botanical Gardens. 2. Huntington
Botanical Gardens–Pictorial works. I. Folsom, James. II.
Bernal, Peggy Park, 1939 . III. Henry E. Huntington Library
and Art Gallery. IV. Title.
QK73.U62H866 1996
580'.74'479493—dc20   95-45328
                                          CIP

FRONTISPIECE: Landscaping at the public entrances—this is one
of two—is a preview of the plants visitors will see in the main
gardens beyond.

PAGES 2–3: The Library setting is tranquil and
elegant, with a broad expanse of luxuriant lawn
in front and the majestic San Gabriel Mountains
in back.

TITLE PAGE: The North Vista, lined with statues, stately palms,
and mounds of azaleas, is a favorite photographic subject
for visitors.

Historic photos are from the Huntington's
photograph collections.
All other photographs copyright © Don Normark,
except as follows:

page 21: Vernon G. Snyder, Aerial Photography
page 50: Walter Houk, copyright © Walter Houk
page 52: Joe Clements
pages 54-55: Robert Schlosser
page 56, top and bottom left: Joe Clements
page 62: Joe Clements
page 63, top: Joe Clements
page 64 top right, bottom left: Joe Clements
page 67, top: Kit Amorn
page 67, bottom: John N. Trager
page 68: John N. Trager
page 69: John N. Trager
page 88: Robert Schlosser
page 96, left: Ann Richardson
page 97: Grady Perigan
page 119: Warren Hill
page 131: John N. Trager
page 134: John N. Trager
page 135: John N. Trager
page 136, top: Kathy Musial
page 152, right top and bottom: Noramelia Martinez
page 153: Noramelia Martinez
page 161: James P. Folsom
page 164, top right: John N. Trager
page 178: Robert Schlosser
page 180, left: Robert Schlosser
page 181: Gerald R. Fredrick
page 182: Robert Schlosser
page 183: Robert Schlosser
page 184: Robert Schlosser
page 185: Robert Schlosser
page 186, column 1, center right: Ann Richardson
page 186, column 2, bottom left: Joe Clements
page 187, column 1, top: James P. Folsom
page 187, column 1, center left and right; column 2, top left
and right: John N. Trager

CONTENTS

Desert Garden

Lily Ponds

Australian Garden

Zen Garden

Subtropical Garden

Jungle Garden

Rose Garden

Japanese House

HUNTINGTON GALLERY

Palm Garden

Patio Restaurant

Japanese Garden

Conservatory

Ikebana House

Shakespeare Garden

Herb Garden

LIBRARY

North Vista

Camelia Garden

PAVILLION

SCOTT GALLERY

Information

Camelia Garden

Deodar Lane

Parking

Mausoleum

Orange Grove

THE HUNTINGTON
Library, Art Collections, Botanical Gardens

A VIEW OF THE HUNTINGTON FROM THE NORTH

Map courtesy of The Times Mirror Foundation

This map plots the public area of
the gardens, which cover about 150
acres; the rest of the property (not
shown) is undeveloped and
not open to the public.

• FOREWORD •

HENRY EDWARDS HUNTINGTON purchased James De Barth Shorb's San Marino Ranch in 1903, flush with enthusiasm for the accommodating climate and the economic future of Southern California. Although he did not arrive with a plan for developing the property, he knew from previous visits to the ranch that this land near Pasadena would support cultivation of oranges and peaches, palms and roses. He brought to the property a passion for development that united his expanding avocations—building great literary, research, art, and plant collections, examining the bountiful horticultural and agricultural potential of the local climate, and creating the structure and resources for his dynamic legacy—The Huntington Library, Art Collections, and Botanical Gardens.

The Huntington Botanical Gardens respond to his vision still today. Even as we approach a new century and face changing times with newly apparent environmental, social, and economic concerns, the values of plants and gardens as well as the importance of botanical, horticultural, and agricultural investigation, remain fundamental and undiminished. Mr. Huntington appreciated the basic need to press ahead with the study and explanation of the nature and importance of plants. As an educational museum and research institution, this garden strives to serve the public welfare through continuing dedication to that mission.

JAMES FOLSOM
Director of the Botanical Gardens

• HISTORIC OVERVIEW •
NATURE'S BEAUTY ENHANCED BY MAN'S DESIGN

SOME VISITORS see the Huntington estate as one of the world's most beautiful gardens. Some see the place as a living museum of the rare and wonderful in the world of plants, a laboratory for the pursuit of botanical and horticultural knowledge. Some see it as a princely repository of books and paintings amid decorous lawns and groves. All the above views are valid. Each sees part of a remarkable synergy of literature, art, and nature that creates a whole far greater than the sum of its parts. This book will focus on the garden part, but first let us take a look at that magical whole.

Founded in 1919 by Henry Edwards Huntington, the Huntington Library, Art Collections, and Botanical Gardens is an educational and cultural center serving both scholars and the general public. Opened to the public in 1928, the year after its owner died, this renowned legacy is an assemblage of collections unique in the world, inviting comparison with any great house in Europe or America.

The Huntington Library is one of the largest and most complete research libraries in America in the fields of British and American history, literature, and science. It contains more than 600,000 books and more than three million manuscripts. Some are part of the changing public display in the Library's exhibition hall, among them such treasures as the Ellesmere manuscript of Chaucer's *Canterbury Tales* (c. 1400), a Gutenberg Bible (c. 1455), and the manuscript of Benjamin Franklin's *Autobiography*, one of the most important documents in American history.

Heart-shaped sprinkler head

LEFT: This pastoral scene of J. De Barth Shorb's San Marino Ranch appeared in *History of Los Angeles County, California*, published in 1880 by Thompson and West. The artist is not identified, but according to the title page, the illustrations were "from original sketches by artists of the highest ability."

Henry Edwards Huntington amassed a fortune as a railroad and real estate developer, then spent some of it assembling major collections of rare books, manuscripts, paintings, and plants.

Arabella Duval Huntington married Henry Huntington in 1913; she was previously married to his uncle Collis, who died in 1900. She and Henry were close in age and had many interests in common.

The art collections are housed in two separate gallery buildings and in the Library's west wing. The Huntington Gallery, originally the estate residence, contains one of America's most distinguished collections of British paintings and sculpture of the eighteenth and early nineteenth centuries, as well as outstanding eighteenth-century French decorative art. Its famed paintings include Thomas Gainsborough's *The Blue Boy*, Sir Thomas Lawrence's *Pinkie,* and Sir Joshua Reynolds's *Sarah Siddons as the Tragic Muse*. The Virginia Steele Scott Gallery of American Art exhibits American paintings from the eighteenth to the twentieth centuries, as well as work by the noted Pasadena architects Charles and Henry Greene and artisans of the early twentieth-century Arts and Crafts movement. A gallery in the Library displays the Arabella Huntington Memorial Collection of Renaissance paintings, including a Roger van der Weyden masterpiece *Madonna and Child*, and eighteenth-century French sculpture and decorative arts.

Surrounding the buildings are 150 acres of gardens with sweeping lawns and vistas, graced by pavilions, pergolas, arbors, fountains, ponds, statuary, decorative tempiettos ("little temples"), and such other structures as the Huntington mausoleum in the tempietto form. The grounds are luxuriant with 20,000 different kinds of plants from the world over, displayed as botanical and landscape subjects in fifteen separate gardens. Among specialized botanical themes are a desert garden with one of the world's largest collections of cacti and other succulents and a garden that shows the history of the rose over two and a half millennia. Elsewhere the estate contains remnants of its working-ranch past and of its earlier native plant cover.

Such diverse collections reflect the extraordinary range of interests of Henry Huntington, a railroad and real estate developer who had enormous impact on Southern California in the early twentieth century. His Pacific Electric Railway, incorporated in 1901, grew into a vast street-rail and interurban rapid transit system, defining patterns of settlement that would structure the future Los Angeles metropolis and indeed predict its freeway network. As his rail lines expanded, he also bought, developed, and sold

Henry E. Huntington and his uncle Collis P. Huntington (right), photographed on a San Francisco street in the 1890s

property along their rights of way to amass the largest landholdings in Southern California and greatly enlarge an inherited fortune.

He achieved most of that in little more than the decade before he retired from business in 1910. Meanwhile, starting in 1903 he devoted nearly two decades to developing the country place we know today simply as "The Huntington."

The complex occupies an astonishing site on an eminence above the San Gabriel Valley east of Los Angeles. Like a stage backdrop, the awesomely steep San Gabriel Mountains, snow-capped in winter, tower to the north. To the south the property commands a sweeping view over the broad valley and on the clearest days to the sea beyond the Los Angeles plain.

The Huntington's gardens grew out of Southern California's golden age of horticulture from the late nineteenth century well into the twentieth. During that time wave after wave of newcomers, captivated by the mild Mediterranean climate, transformed this favored land in a tour de force of landscaping with plant material from the world over. While sharing their fascination, Henry Huntington was aware that established gardens all around had been planted with nineteenth-century exuberance and lack of discipline. In developing his own estate he too would use new and exotic material but he added the dimension of planting for systematic study. His collections aimed to yield information, ever enlarging the vocabulary of landscape art while increasing scientific knowledge. Today his gardens still flourish, one of the few surviving examples from that glorious, optimistic era. And now, as then, certainly the finest.

BORN IN 1850 in Oneonta, New York, Henry Edwards Huntington left home at age twenty to work for a New York hardware supplier. Before long his uncle, railroad baron Collis P. Huntington, recruited him to manage a West Virginia railroad-tie sawmill. Ten years later Collis appointed him construction superintendent of the Chesapeake, Ohio and South Western Railroad in Kentucky and in a few years made him manager of the

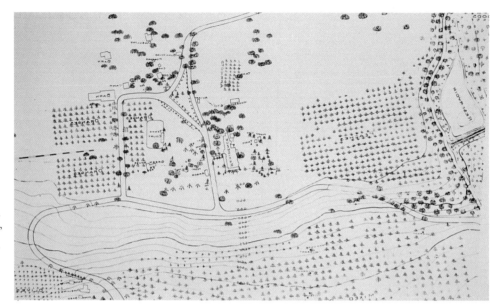

This ground plan of San Marino Ranch in 1904 shows areas devoted to oranges, peaches, apricots, pepper trees, and a plot described as "mixed orchard." Existing structures, including a residence, barns, bunkhouse, and tanks, are also indicated.

Kentucky Central. By 1890 the railroad world recognized two Huntingtons, calling the younger "H.E." to distinguish him from the more famous "C.P."

In the spring of 1892 Henry made his first trip to California. He was headed for the San Francisco headquarters of the Southern Pacific Company, the SP, the powerful holding company for the Southern Pacific and Central Pacific railroads. Henry would be special assistant to the SP president, Collis Huntington, looking out for his uncle's interests in the West while Collis spent most of his time on other business in the East. Traveling in C.P.'s private railroad car, the two Huntingtons took the Southern Pacific Railroad, southernmost of the transcontinental lines, giving H.E. his first look at that operation. A stopover in Los Angeles gave him his first glimpse of Southern California.

Henry had a knowledgeable guide, for Collis had lived in California since the pivotal event in the state's history, the 1849 Gold Rush. The general merchandise business he operated in Sacramento with partner Mark Hopkins had become one of the West's largest and most prosperous by the time they joined two other merchants, Leland Stanford and Charles Crocker, in a new enterprise. Soon known as the Big Four, the partners won both fame and notoriety for financing and building the Central Pacific Railroad in 1869. California's first transcontinental rail connection, it later became part of the Southern Pacific system.

DURING THE DECADE BEFORE H.E.'s visit, Southern California had experienced a phenomenal boom fueled in part by a railroad fare war. That event marked the end of a monopoly that Southern Pacific had enjoyed from the time it reached Los Angeles in 1876 to become the city's sole rail link to the East. In 1885 a second railroad, the Atchison, Topeka and Santa Fe, arrived and competition between the two quickly grew intense. For nearly a year a passenger ticket from Kansas City to

Turn-of-the-century view of Pasadena, a city described in a contemporary publication as one of "magnificent homes, schools and hotels and no bars. . .Pacific Electric cars every five minutes."

Los Angeles cost less than $25, one day briefly dropping as low as one dollar. All the while a flood of booster literature trumpeted Southern California as a paradise with a healthy climate, warm winters, and excellent growing conditions. A typical panegyric called the northwestern San Gabriel Valley "a garden of Eden where grew every tree that is pleasant to the sight and good for food." A more level-headed chronicler noted such drawbacks as rainy days, hot spells in summer, violent winds, even earthquakes. But no voice of reality could compete with the promoters' siren song, and migrants from east of the Rockies continued to pour in by the trainload.

Through his uncle and on his own, Henry was an insider at Southern Pacific, which was a prime mover of the boom. Not just a railroad, SP was also a major landowner, a redoubtable player in politics, local and national, and an active publicist of destinations along its routes. In 1898 it even founded a magazine, *Sunset*, named for its crack Sunset Limited passenger train, to extol the new land's wonders. Meanwhile Uncle Collis had spent much time furthering the railroad's land and route acquisitions and working in a failed attempt to have Los Angeles locate its new harbor in Santa Monica—where SP would have been in control.

His railroad and harbor activities gave C.P. a wide acquaintance among Los Angeles bankers, politicians, merchants, and ranchers. One landowner, James De Barth Shorb, extended his hospitality during the Huntingtons' brief stay in 1892. Their visit to the Shorb ranch was H.E.'s first brush with his future estate.

History mostly remembers Shorb as the son-in-law of Benjamin Davis Wilson, a Yankee pioneer who became a major landholder, businessman, and first mayor of Los Angeles. Among Wilson's properties was one called Lake Vineyard, seven miles northeast of Los Angeles and fertile with grapes, oranges, olives, figs, and half a dozen other orchard crops. He had given Shorb part of it, called Mount Vineyard, and under Shorb's stewardship both became well known. An 1874 book

Known as the Big Red Cars, the trolleys of Pacific Electric handled thousands of commuters every day and offered such special sightseeing trips as the one advertised in this brochure for a visit to the "Orange Kingdom."

details the two estates' walnut and citrus groves, more than a million oranges shipped in one season, 1,500 acres in grain, and 90,000 gallons of cellared wine.

Shorb renamed his ranch "San Marino" after his boyhood home in Maryland, and built his house at the brow of a prominent slope, actually a long east-west fault scarp that provided the San Gabriel Valley some of its finest view sites. One writer described the oak woodland to the north as a magnificent natural park and noted an expansive vista south to Santa Catalina Island. Another writer called Shorb's mansion the finest in Pasadena (the nearby town), adding that it was surrounded by tropical vegetation. The future Huntington estate was off to a promising start.

IN THOSE HEADY DAYS, folks from cooler climates bought land in this new garden spot and eagerly planted natives and a gamut of imports from tropicals to tundra. One tenderfoot thought this must be enchanted ground, compared with New Hampshire's nine months of winter and three of pretty cold weather. Many were the wonders to write home about. In the year-round season, plants grew larger and faster than any the newcomers had seen. Tomatoes attained record size and bore fruit in winter. Vines shot up like Jack's beanstalk. Geraniums that could barely make it through winter in a Boston window box here grew into hedges of great height.

By the turn of the century Southern California was full of plant enthusiasts from the East and Europe. Horticulturists and hobby gardeners, nursery growers and landscape designers—all were happily trying out what would thrive here. The University of California undertook field research. The U.S. Department of Agriculture joined in the worldwide search, bringing back test material for its own experiment stations and for distribution to individuals, nurseries, park departments, arboretums, and gardens. In time it would give more than five thousand plants to the Huntington.

So exuberant was the planting binge in the San Gabriel Valley that guidebooks and *Sunset Magazine* directed coach-borne tourists to truly parklike ranches and estates, promising a rich bounty of sightseeing pleasure. The Shorb and Wilson properties were prominent features of tours. So was Elias J. "Lucky" Baldwin's Rancho Santa Anita, with its thousands of trees, shrubs, vines, and flowers—now the Arboretum of Los Angeles County. Tours might visit Carmelita, the more modest Pasadena garden planted by Jeanne C. Carr with cuttings and seeds from the nearby San Gabriel Mountains or brought from their travels by the celebrated naturalist John Muir and other friends. The nearby Busch Gardens, the estate of St. Louis brewer

Adolphus Busch, offered miles of paths through thirty acres of gardens full of statues from the Grimm fairy tales brought from Germany—until the gardens closed during World War I amid mounting anti-German sentiment. Also nearby was Kinneloa, the foothill ranch of State Forester Abbott Kinney, famous then for promoting eucalyptus planting, remembered now for his Venice, a canal development beside Santa Monica Bay. At General George Stoneman's place, visitors marveled at plantings of stone fruits and such exotics as bananas, guavas, pomegranates, tamarind, and a 20-foot-high 'Lemarque' rose that produced more than 180,000 blossoms at a time. Or so it was said.

Tourists took the so-called Inside Track, a Southern Pacific line east across the San Gabriel Valley and beyond as far as Redlands. There they visited such showplaces as the renowned Cañon Crest, a residential park developed by the brothers A.H. and A.K. Smiley. Extravagantly proclaimed to be unequaled under Italy's sunny skies or amid Switzerland's alpine wonders, its 200 acres of flowers, rare shrubs, and subtropical trees were laced by five miles of road.

Pasadena, the San Gabriel Valley's most elegant community, also attracted visitors. Founded as a citrus growers' town, it grew in the boom of the 1880s into a winter resort for the wealthy and cultivated, with fashionable hotels, opera house, public library, and other amenities. Many well-heeled Pasadena citizens had come West for a winter escape, liked what they saw, and came back to build houses, even if only for winter occupancy.

The bounty of a land that seemed perpetually in bloom inspired annual flower festivities. Los Angeles put on a parade of flower-bedecked coaches, tally-hos, automobiles, fire engines, and floats past store windows filled with flowers. Riverside had a street fair and its own floral parade. Santa Barbara had a floral festival that included an event in which normally sane citizens pelted one another in a jolly Battle of Flowers. Pasadena staged what came to be the finest pageant of all and the only one to endure. From 1890 on its Tournament of Roses has made New Year's Day a celebration of everyone's good fortune to be in the land of sunshine rather than buried under snow back East.

Henry Huntington rests beside an oak saved by tree surgery, its cavities filled with steel-reinforced concrete. William Hertrich wrote, "There was no subject Mr. Huntington loved to discuss. . . more than the trees he had been able to save."

Henry Huntington was as charmed as anyone. He thought the San Gabriel Valley the prettiest place he had ever seen, especially in springtime. Here he, too, would eventually build a winter retreat of his own.

AFTER THAT NIGHT with the Shorbs in 1892, Henry Huntington went on to San Francisco and his position as a high official at Southern Pacific. With his wife and four children, he led an active and socially prominent life for several years. When C.P. died in 1900, H.E. inherited a fortune, but his hope of succeeding his uncle as company president was thwarted when he was outvoted by stockholders. He decided to withdraw from management.

Thus at the turn of the century Huntington turned his attention to Southern California, which he accurately saw as a land of promise. Here his major business accomplishment would be a rail empire he had begun in 1898 with the acquisition of the Los Angeles Railway, the LARY. He soon acquired other lines in and around Los Angeles, eventually integrating a hundred traction companies into the Pacific Electric, the PE, the largest interurban network in the world. In its heyday, PE's literature boasted of 2,700 scheduled trolleys and trains a day and its standard-gauge, horn-blowing Big Red Cars rumbled over more than a thousand miles of track. Linking fifty communities, PE rails ran from central Los Angeles east to Redlands, southeast to Newport Beach, south to Los Angeles Harbor, west to Santa Monica Bay towns and beaches, northwest into the San Fernando Valley, northeast to the San Gabriel Valley—and of course San Marino.

Where the PE crossed Los Angeles on city streets, it used trolleys smaller and slower than the big interurbans. On some routes they competed with the bell-clanging yellow cars of the LARY. Since its narrow-gauge track could not connect

William Hertrich, superintendent of the Huntington Botanical Gardens from 1905 to 1948, examining the blossoms of a camellia. Hertrich authored several books and articles on plants, including a three-volume work on camellias in the Huntington gardens. He was a prominent authority and received numerous awards from national and international organizations for his horticultural contributions.

RIGHT: Hertrich became an expert at transplanting large trees because Henry Huntington wanted the gardens to look mature as quickly as possible. These workers are transplanting a three-ton cactus, *Cereus alacriportanus*, in 1925.

with PE's wider standard gauge, Huntington operated the LARY as a separate street rail system, considered in its day to be the world's best.

Huntington built power generating and other utilities for his electric railways and for the real estate that he coordinated with his rail systems. Knowing well in advance where PE lines would be going gave him an inestimable advantage in buying up land for later subdivision. He left his name all over the map, on the towns of Huntington Beach and Huntington Park, on such districts as Huntington Palisades and Huntington Riviera, on Huntington Drive through San Marino, and the Huntington Hotel and Huntington Hospital in Pasadena.

Moving to Los Angeles in 1902, he first lived in a hotel, then moved into the exclusive Jonathan Club, both downtown. He had left his family, the children grown, in San Francisco, and his marriage ended in divorce a few years later. In 1903 he purchased the 600-acre San Marino Ranch from the bank that had assumed ownership after the death of his friend James Shorb. There in a few years he would build a new house where the Shorb mansion stood, but the place had fallen on hard times and rethinking the landscaping was the first order of business.

AFTER THE FIRST MAN he hired for the project failed, Huntington was fortunate to find William Hertrich, not only an unusually able landscape gardener but a kindred spirit in his approach to horticulture and the land. The German-born Hertrich's professional experience had begun at age sixteen with a four-year apprenticeship in an Austrian firm, then he went on to courses in agriculture, horticulture, landscape gardening, and estate management. Emigrating to America in 1900, he first settled in Connecticut, then moved to California, where he had an uncle living in Orange County. After several landscape jobs nearby, he hired on at San Marino Ranch in

1921: An aerial view of the north section of the estate when it was still occupied by the Huntingtons. The 600-acre ranch was reduced to 207 acres when it became a public institution.

1904. In letters and in his book, *The Huntington Botanical Gardens, 1905–1949, Personal Recollections of William Hertrich*, he recounts the development of the estate and his working relationship with Henry Huntington.

Concern with what today we would call infrastructure indicates that Hertrich's employer had a large vision for the premises as well as the budget to carry it out. Their first tasks were to study the lay of the land, upgrade existing facilities, and get roads, irrigation, and fencing in place. They also established a nursery for grand-scale planting to come. Special gardens within the larger framework would follow in later years.

One of Huntington's landscaping tenets was to avoid formality where possible, except for several spaces designed later on to relate to the house and library. He considered his handsome native Engelmann and coast live oaks, of which perhaps a thousand still remain, an asset to be cherished. For preservation, some oaks were transplanted, and new planting had to respect the positions of those left in place.

In selecting plant material Huntington wanted to emphasize flowers for color in winter and early spring, the time of year when he intended his house to be occupied. The annual bloom calendar still shows the most kinds of plants flowering in February through April.

Proving to be a practical rancher, H.E. raised oranges as a cash crop. He started the first commercial avocado orchard in California with 300 seeds, many of them saved for him by the chef at the Jonathan Club in Los Angeles. Seeking practical information useful to a major land and rail developer, he collected trees, particularly palms, and other plants to test their adaptability to home gardens, parks, streets, and rail rights of way.

Since Huntington wanted his grounds to have an established look as quickly as possible, Hertrich looked for mature as well as rare specimens as he drove about with

Today: The Huntington mansion (today the Art Gallery) is seen in the foreground, flanked by the Library building and entrance pavilion on the right and the Virginia Steele Scott Gallery on the left. The city of Pasadena sprawls north toward the San Gabriel Mountains.

his horse and buggy, even on Sundays, cruising nurseries and other sources. He would truck in large trees from near and far, often under difficult conditions, and even brought three carloads of young saguaro cacti by train from Arizona.

Huntington and Hertrich would also travel together seeking plant material and lore at nurseries and other estates from San Diego to Santa Barbara. They visited some legendary plant specialists.

In San Diego they met with nursery owner Kate Sessions, noted for introducing trees and plants from Mediterranean, tropical, and desert climates and for using drought-tolerant California natives. In Santa Barbara and Montecito they likely visited gardens with such names as Ceanothus, Green Acre, and Take It Easy. They surely called at the nursery of the renowned Francesco Franceschi, an Italian immigrant credited with introducing more plants into California than anyone else. And they would have called on Joseph Sexton, whose long-established nursery was virtually a botanic garden.

East of Los Angeles, the Armstrong Nursery in Ontario interested them for its

The three-story Victorian mansion of James De Barth Shorb was a mecca for prominent visitors, including Henry Huntington, who was a guest there in 1892. Huntington built his own mansion on the same site eighteen years later.

experiments with fruit and ornamental trees, while the city's street plantings of eucalyptus, pepper, grevillea, and palms were already attracting attention. In Los Angeles the two called on Theodore Payne, who offered America's most complete stock in seeds of palms and other trees, shrubs, and California wildflowers. They visited Howard & Smith, with its huge collection of flowering plants; Hugh Evans, specialist in rare and esoteric introductions; and E.D. Sturtevant, the foremost authority on water lilies.

In 1904 the Sturtevant nursery supplied the water lilies, lotus, and other aquatic plants for the first special garden at San Marino Ranch, a set of lily ponds started by his predecessor but completed by Hertrich. Most water lilies and lotus bloom in summer, but this installation provided warm water so that spectacular giant Amazon water lilies could put forth flowers outdoors until mid-winter. That first for California was an impossibility in most of America.

Two of the most specialized collections, palms and desert plants, began the next year. When Hertrich proposed the first, Huntington readily agreed, for palms were a personal favorite. They began bringing in species from the tropics and from places with temperate climates similar to that of San Marino to test their local performance, a procedure they would follow again and again when garden developments came up for consideration. Hertrich found many specimens in nearby nurseries, traveled east to find others, and had some sent from nurseries in Europe and Japan.

Huntington was less enthusiastic about desert plants, remembering a prickly encounter while supervising SP rail building in Arizona. Yet little-known desert plants seemed the only solution for a prominent area of poor soil that defied conventional greenery. At last the astute Hertrich won approval by appealing to his employer's collector instincts. A cactus and succulent collection evolved gradually from the few species Hertrich could find locally, augmented later with shipments from Arizona and Mexico. Eventually enlarged to twelve acres, the desert garden achieved such stature that some former competitors for specimens donated their collections to the Huntington. In our time it shines as a star of the show.

Other installations followed. By 1909 a greenhouse for orchids and other tropicals had been built, a collection of camellias was under way, and a collection of

half the known species of cycad was in place. A Japanese garden, now one of the most popular features of the estate, was completed in 1912.

With landscaping well along, the time came to build the house. Huntington selected the well-known architectural firm of Myron Hunt and Elmer Grey to design a stately new mansion in the Beaux Arts style on the site of Shorb's old Victorian house. After a year of construction the residence—today's Huntington Art Gallery—was completed in 1911. Its related outdoor spaces included an impressive loggia, a broad terrace, and two formal gardens—a rose garden to the west and the North Vista. This long grass allée, lined with a living colonnade of smooth palm trunks, focused on a baroque fountain backed by dramatic mountains beyond. Later on, Italian garden statuary added to the palms' architectural effect and camellias and azaleas provided winter color.

BY 1910 HUNTINGTON had retired and put aside most business affairs to devote his time to art and book collecting. He already had a formidable reputation based on his avid and increasingly knowledgeable pursuit of literary material, often buying rare manuscripts, volumes, and whole libraries at or even before auction. In time he would become America's foremost book collector. Eventually his holdings grew so great that he engaged Myron Hunt to design another major building, a library, on his estate grounds. Begun in 1919, it was completed the following year, though the interior would require several more years to finish.

In 1913 he married the widow of his uncle Collis, Arabella Duval Huntington, who was close to his age. One of America's wealthiest women, especially after Collis's legacy, she was also one of the important art collectors of her time. Sharing Henry's interest in art, she advised him on some purchases as his own collection grew ever more consequential. The newlyweds moved into their palatial new house in 1914. Until Arabella died in 1924 they divided their time between San Marino and their New York mansion, with an occasional stay in Europe.

Huntington was fully aware of the stature of his estate. In 1919 he assured its

A large domelike cage in the aviary's center housed twenty-six kinds of parrots and sixty-five other kinds of seed-eating birds. Built for Arabella Huntington, who was fond of birds and animals, the aviary had twelve sections for different special collections. It was torn down after her death.

LEFT: This view of the Huntington mansion was the frontispiece for the 1915 book *Stately Homes of California* by Porter Garnett. He described the house as "the most 'palatial' residence in California" and the gardens "in outdoor plants, among the richest in the world."

future with a trust indenture to create and fund the library, gallery, and gardens that would grow into today's institution, on 207 acres of the original premises. Before the ranch was reduced to that size, H.E. proudly pointed out that its production of citrus and other crops had made it self-supporting. The new boundaries preserved about 20 acres of orange groves along with the best of the ornamental plantings.

Huntington lived until 1927, spending most of his final years in San Marino. As the end neared, he had architect John Russell Pope design a mausoleum where he and Arabella would be interred. She had selected the site at the property's high point, in the northwest section by the orange groves. A prototype for Pope's future Jefferson Memorial in Washington, the chaste white marble structure was completed in 1929, after the death of its occupants.

In Huntington's lifetime the garden collections expanded through propagation, purchase, trading, and donation of plants, rootstock, and cuttings from nurserymen and other sources. Though funds for acquisition dropped after 1927, gifts continued, mainly because of William Hertrich's wide personal contact with plantsmen. Hertrich retired in 1948, but continued to watch over his beloved gardens until, frail at 88, he died in 1966.

WHEN HENRY HUNTINGTON passed away, half the 207 acres had been developed as gardens. Today that figure has grown to about 150 acres landscaped and open to visitors.

Preparation for opening the estate to the public required such changes as creation of a parking lot and labeling plants. The first visitors were admitted to a fifty-acre tract around the mansion and Library and the Japanese and desert gardens, the maximum area that could be adequately guarded. Security also required the number of visitors to be limited through a reservations system. Over the years other parts of the gardens would open in increments.

Plant acquisitions continued. In the decade following 1943, three major additions, including a gift of 450 trees from Los Angeles County, increased the collection of cone-bearing trees—pines, cedars, cypresses, and such Southern Hemisphere trees as kauris and araucarias—to one of the largest in Southern California.

The first version of the Shakespeare garden, designed in 1959 by landscape architect Ralph Cornell, featured an English sundial, a sculptured bust of Shakespeare, and plants arranged in beds separated by paved walks.

During World War II, more than half the garden staff joined the armed forces and maintenance declined. A reduced labor force also meant that no major new developments were undertaken. Meanwhile, San Marino Garden Club members raised some 40,000 plants on the grounds for the U.S. Army Corps of Engineers to use for camouflage. At the same time an intensive effort was made to propagate *Aloe vera* for use by hospitals as a burn treatment.

New garden works resumed in postwar years. In the 1950s what had been Arabella Huntington's cut-flower garden was converted to an herb garden, then in 1976 that was completely rebuilt and replanted in its present form. Also in the 1950s, a camellia garden officially opened to the public, culminating planting and breeding efforts Hertrich had begun forty years earlier.

A garden devoted to plants mentioned in the plays and poetry of Shakespeare was opened in 1959. A direct reference to the Library's considerable holdings of the Bard's works, this garden held the likes of hawthorn, holly, laurel, and yew among more than a hundred kinds of plants. It also provided a link that for the first time let visitors walk on garden paths and a terrace completely around the Huntington Gallery. In 1984 the Shakespeare garden was redesigned and replanted, with the new Scott Gallery as background.

Like the camellia collection, an Australian garden emerged from plantings made much earlier. Begun in 1943 as a test plot for eucalyptus, it was enlarged with other Australian material and formally established as a separate unit in 1964. Plants added since then have been mostly Australian species, including those used as floral emblems by that country's six states and two territories. In 1987 paths were installed to improve public access.

In 1968 the most ambitious project in decades doubled the Japanese garden in size with a gravel and rock garden in the Zen style, a bonsai court, and other features.

Subtropical plants had been used throughout the gardens, but until 1964 no area was specifically devoted to their cultivation. Planting of subtropicals began that year on a four-acre, south-facing slope below the Huntington Gallery and above the Australian garden. The site promised protection for frost-tender species, since cold air flows downward and is less likely to injure plants on a slope than below at a lower elevation. The section opened to the public in 1976, after specimens had matured and paths had been installed.

The jungle garden was another for which planting had begun years before it officially opened to the public in 1979. On the slope between the subtropical and palm gardens, it has luxuriant foliage in contrast with the nearby desert garden, around a cascading waterfall as centerpiece.

Also that year, a new parking lot opened on the site of Huntington's avocado orchard, its trees long since grown too tall for easy fruit harvesting. Eucalyptus and other exotics earlier planted for testing provided background trees and islands of verdure to soften the effect of paving and create an inviting atmosphere. At its south end the parking lot led to a handsome entrance pavilion, opened in 1981, completing facilities for welcoming visitors.

A desert garden conservatory, a haven for the display and propagation of succulent plants that would not survive winter frost, opened to the public in 1985.

Superintendent Hertrich's successor today is the director of the botanical gardens, one of the institution's major divisions. A staff for research, development, and maintenance includes curators in charge of separate garden areas and plant specialties. Collection and study of plants continue with such activities as field trips, exchange of seeds and plants, propagation, hybridizing, operation of a herbarium, and others.

These days the gardens have many distinctions: one of the world's foremost collections of cacti and other desert plants, one of America's largest collections of temperate-climate palms, the nation's largest collection of camellias, a garden of 1,800 kinds of roses grouped historically, a major Japanese garden with a traditional house and a Zen garden, and many landmark, unusual, and rare plant specimens.

Development of the gardens continues. Botanical and horticultural knowledge are still the primary thrust, the serious part of the gardens' dual importance, while many of the half-million visitors a year come for the aesthetic pleasures of the landscape and the treasures of art and literature they surround. If Henry Huntington were to return, he likely would agree that the gardens contribute the kind of setting one of the world's most important libraries and one of America's most important art collections deserve. The triad constitutes Southern California's premier showplace of the arts of civilization.

OVERLEAF: Huntington loved the oaks on his property. He had 650 of them transplanted from a canyon area so they could be incorporated into his landscaping.

In The Gardens, Part I

• THE LILY PONDS •
OLDEST OF THE SPECIALTY GARDENS

The lily ponds were the first landscape feature to be completed. Here the sacred lotus (*Nelumbo nucifera*) unfurls immense pink-tipped petals above broad upturned-umbrella leaves.

HENRY HUNTINGTON began his first themed garden in 1904, a set of half a dozen lily ponds installed for the practical purpose of replacing a shallow, unsightly gully in the southeast corner of the gardens with something more presentable. To complete the project, begun by his predecessor before he was fired, William Hertrich devised an innovative warm-water system—later abandoned—for one pond that kept such species as the giant water lily *Victoria amazonica* in bloom well into winter. Appropriately, here where it all started a plaque now memorializes Hertrich as the botanical gardens' innovative designer and developer.

Around the ponds several acres without visible boundaries present a vision of Elysian fields, matured over the years into a display of water lilies, lotus, and other aquatic plants in a sylvan setting of bamboo groves, lawn, and tall trees. A cascading stream down from an adjacent jungle garden enhances the scene and through aeration contributes to water quality. The ponds are home to bright-colored koi (Japanese carp), turtles, frogs, and resident and visiting ducks.

Descendants of a long history of water lily (*Nymphaea*) cultivation are seen in the ponds, where many varieties grow in profusion and blossom in varied hues mid-spring through mid-autumn. Their magnificent relative, the sacred or East Indian lotus (*Nelumbo nucifera*), dominates the lowest pond, where the original plants remain from their introduction in 1905. In summer they lift impressive pink

A 1916 view of one of the lily ponds with the mansion at the top of the hill.

and white flowers eight or nine inches across on stems above inverted-parasol leaves fourteen or more inches across.

Along the shoreline cannas, strelitzias, and cordyline contribute bold leaf patterns, and the water-loving papyrus (*Cyperus papyrus*) adds grace. The bulrush of the Bible, this tall sedge used for writing paper in the ancient world grows along the Nile from Egypt south into Uganda.

Close by across a stretch of lawn west of the ponds, the Huntington's unusual conifer collection was planted in this cool spot to favor species from temperate areas. The grove holds several araucarias, Southern Hemisphere natives whose common names include the word "pines," though true pines all come from the Northern Hemisphere. The best-known here is Norfolk Island pine (*Araucaria heterophylla*) and there are specimens of New Caledonia pine (*A. columnaris*) and hoop pine (*A. cunninghamii*). An unusual pine relative, *Keteleeria davidiana,* comes from China. Two trees are cypress species, the rare Guadalupe cypress (*Cupressus guadalupensis*) from the island of that name off the coast of Mexico and the elegant Kashmir cypress (*C. cashmeriana*), one of a species probably extinct in the wild. Two are cypress relatives, Taiwania (*Taiwania cryptomerioides*) from Taiwan and China and Montezuma cypress (*Taxodium mucronatum*) from Mexico. Two are so-called living fossils from that ancient time before flowering plants evolved, the dawn redwood (*Metasequoia glyptostroboides*), a deciduous conifer thought to be extinct until it was discovered growing in China in 1948, and the deciduous maidenhair tree (*Ginkgo biloba*), related to the conifers and extremely rare in the wild.

Among notable non-conifers growing nearby are such flowering trees as strawberry snowball (*Dombeya cacuminum*) from Madagascar, a sweetshade (*Hymenosporum flavum*) from Australia, the largest southern magnolia (*Magnolia grandiflora*) specimen on the grounds—and a specimen of a commercially important food plant from Australia, *Macadamia integrifolia,* beloved by resident squirrels for its late-summer bounty of nuts.

PLANT LANGUAGE

MOST BUT NOT ALL garden plants have common names, which may vary by region. Botanical names, consisting of a genus and species name, sometimes followed by a subspecies, variety, cultivar, or other designation, provide a more standardized terminology. Common names are not italicized, botanical names are in Latin and are italicized. The first word in a botanical name gives the genus (plural: genera), a major subdivision of a plant family that may contain one or many species distinct from one another but closely related. The second word gives the species, a further subdivision in classification. A third word may designate subspecies or variety, plants that share major characteristics with minor differences, or cultivar, a contraction of "cultivated variety" denoting a variety originated in cultivation. In abbreviations, "ssp." means subspecies, "spp." means species in the plural, "var." means variety. The lowest subdivision of a species is forma, abbreviated "f.," used to indicate a trivial variation such as a different color. A hybrid, the result of a cross between two genera, species or varieties, is designated by an "x" between the names, as in *Coleus* x *hybridus.*

Thus the plant name Rosemary (*Rosmarinus officinalis* 'Lockwood de Forest') gives the common name, genus *Rosmarinus,* species *officinalis*, and cultivar name, 'Lockwood de Forest'. If another rosemary is mentioned immediately, the genus would be abbreviated "*R.*" followed by the species and variety.

Some other special terms: endemic, a plant native only to a particular location; epiphyte, a plant that grows without soil on another plant without receiving nourishment from it; exotic, often loosely equated to strange, in botanical usage it is reserved for plants of foreign or out-of-state origin; sport, considered a mutation, a plant whose leaves or flowers vary spontaneously from the normal; xerophyte, a plant adapted to arid conditions; hardy, resistant to frost, and tender, not resistant to frost; succulent, a plant that stores water and nutrients in leaves, stem, or roots.

Water lilies (*Nymphaea*), left and top, are today largely hybrids of some thirty-five species from all parts of the world. The beauty of their shining saucerlike foliage is best seen if there is enough open water for the flowers to float free and be reflected. When its petals fall, the conical seed pod of the lotus, bottom, is revealed as a structural curiosity.

ABOVE: The visitor moves from the lily ponds into the world of bamboo, where tall stands of *Bambusa tuldoides* and *B. beecheyana* tower over the stone bridge.

LEFT: Bright-colored koi (Japanese carp) share the lily ponds with turtles. Breeding has emphasized certain markings on the fish, which are particularly beautiful when viewed from above.

OPPOSITE: The upper lily pond is virtually a secret garden where the sound and movement of a waterfall animate the quiet scene. Water lilies cover the surface of the pool, surrounded by such dense greenery as the huge fringed leaves of *Philodendron* 'Evansii' and the striking fronds of saw palmettos (*Serenoa repens*).

• THE BAMBOO COLLECTION •
THE GIANT GRASSES, LARGE AND SMALL

LIKE PALMS AND BANANAS, bamboos contribute a tropical look wherever they are planted on the grounds, notably around the lily ponds, in the jungle and Japanese gardens, and in the public parking lot. Fifty species of these giant grasses grow in the gardens, with more in areas out of public view. Most are of the clump-forming *Bambusa* genus, whose rhizomes or underground stems grow out only a short distance before sending up new vertical stems, and the running *Phyllostachys* genus, whose rhizomes grow for greater distances before sending up new stems.

Two large stands of punting-pole bamboo (*Bambusa tuldoides*) and *B. beecheyana*, tallest of the clumping bamboos, are prominent just west of the lower lily ponds. On the east side and among the ponds, dense groves meet overhead to create shady tunnels over water and pathways. These are *Phyllostachys flexuosa, Bambusa glaucescens,* and *B. oldhamii*—a giant timber bamboo specimen planted in 1906 as one of Henry Huntington's first garden acquisitions. Isolated in the lawn near the middle pond is a stately B. *tulda* with thick, bluish stems. North of the ponds is a clump of Buddha's belly bamboo (*B. ventricosa*), a curiosity that under drought stress produces inflated internodes—the sections of bamboo canes between joints or nodes—that suggest the ample belly of Buddha statues.

In the jungle garden by the waterfall is the rare *Dendrocalamus asper* from south Asia, a member of the tallest-growing of the forty-five genera of bamboo. It has dense tan fuzz on young stalks and handsome foliage, and can grow fifty feet tall. Close to the jungle garden's eastern edge grows one of the small clumping bamboos of the American tropics, *Chusquea coronalis*, a rarity that has long delicate green plumes when in active growth.

In the Asian plant section of the Japanese garden luxuriant stands of golden or fishpole bamboo (*Phyllostachys aurea*), one of the best kinds for edible shoots, line both sides of the south or lower exit path. Two others nearby are black bamboo (*P. nigra*), named for its dark stalks under a mass of short, light green leaves, and the large-caned *P. vivax*.

The parking lot outside the entrance pavilion is the setting for one of the gardens' most beautiful bamboos, *Bambusa vulgaris* 'Vittata', a striking form with thick yellow stalks striped in panels of dark green.

OPPOSITE: To the east side of the lower lily ponds is a large grove of bamboo. Its tall green stalks and leafy canopy far overhead filter the light, creating a cool dimness. Three kinds of bamboo grow here: *Phyllostachys flexuosa, Bambusa glaucescens,* and B. *oldhamii.*

New spring shoots of giant timber bamboo (*Bambusa oldhamii*) poke up through a mulch of its own leaves in the Japanese garden.

OPPOSITE: An impenetrable stand of golden bamboo (*Phyllostachys aurea*), its slender stems topped by a cascade of pale green, guards the south entrance to the Japanese garden.

The Palm Garden
Icons of Tropical Climes

Huntington's first collection garden, devoted to palms, tested many tropical species that did not survive severe winter frosts. Today the emphasis is on temperate-climate palms.

LEFT: The pleated leaves of a low-growing, widely branching Mediterranean fan palm (*Chamaerops humilis*), the only palm native to Europe, sweep the ground, revealing the dense panicles of yellow flowers at its heart.

RIGHT: The fruit and flower clusters of the solitary fish-tail palm (*Caryota urens*) can grow ten feet long.

BEYOND THE LAWN south of the entrance pavilion a grove of palms, now dense, now more open, creates one of the Huntington's most remarkable landscapes. Here four acres of the most decorative and botanically interesting temperate-climate palms form the nucleus of a continually enlarging collection of more than ninety species.

Palms were favorite trees of Henry Huntington's, by themselves and as landscaping subjects in gardens and public places and along streets and rail routes. Indeed they were immensely popular in Southern California of the Golden Age for their evocation of the tropics and the romance of faraway places. Some species also came to be known as the most architectural of trees, in lines along stately avenues and in formal landscaping. In time the palm became the region's signature tree.

Begun in 1905, the Huntington collection grew quickly with many large specimens and many tropical species. The tender species provided the kind of knowledge expected from test plantings: they perished in frosts that occurred once each decade from 1913 to 1949. It became evident that not all palms that thrive close to the coast are hardy in the San Gabriel Valley's colder winters, yet some early-day veterans remain and new introductions are being tried, mostly in a warm

Outside the palm garden, Guadalupe Island palms (*Brahea edulis*) line Mausoleum Road. Their handsome, regular crowns and stature make them useful avenue trees.

micro-climate shared with the adjacent jungle garden (page 147).

A walk through this small forest reveals a strong family resemblance among palms, a close look turns up striking variations. Most have single upright, unbranched trunks but some have multiple trunks. Most have fan-shaped or feathery leaves atop their trunks but some have undivided leaves and some grow fronds close to the ground. Some reach great height, a few are small. And though most palms prefer the tropics and subtropics, some grow in colder latitudes. Here are a few of special interest:

The Mediterranean fan palm (*Chamaerops humilis*), the only one native to Europe, gave palms their name. Noting a resemblance to the human hand's palm and fingers, the Romans called the tree *palma*, the Latin root of the English word. The garden's large old specimens vary in habit from individuals to clumps, from four-foot shrubs to twenty-foot trees.

The Chilean wine palm (*Jubaea chilensis*), endemic to the valley of Ocoa in central Chile, is the most massive of the palms, with a three-foot-thick trunk that resembles a stocky column from an early Greek temple. It is prized for its sugary sap, which ferments into a crude wine but which can be extracted only after the tree is felled, so many trees in native stands have been lost and the species is endangered.

Remaining from a planting along the estate's former main access road are South American jelly palms (*Butia capitata*), with stout trunks and feathery, gray-

This clump of trunks belongs to the date palm (*Phoenix dactylifera*), which is usually grown commercially as a single-trunk tree. Shrimp plant (*Justicia brandegeana* 'Chartreuse') surrounds the palms with color.

green leaves. In the fall hundreds of their edible yellow or orange fruits dot the lawn.

Several of the date-palm genus *Phoenix* adapt well here. The familiar Canary Island date palm (*P. canariensis*) is California's most widely planted large street and garden palm. A stately specimen in the nearby desert garden was shipped by Henry Huntington from his uncle Collis P. Huntington's house in San Francisco, destroyed in the 1906 earthquake and fire. The date palm (*P. dactylifera*) also grows here, but bears fruit only in hot desert locations. The cliff date palm (*P. rupicola*), from rugged places in Sikkim in the Himalayas and Assam in India, is represented by three specimens twenty-five feet tall, with slender trunks and bright green leaf crowns.

The Canary Island date palm is magnificent for its imposing mature height (to sixty feet) and spread of its crown (to fifty feet). The majesty palm (*Ravenea rivularis*), of which a young specimen is seen here, merits its common name for its perfection of form. A rare windmill palm from the Himalayas, *Trachycarpus martianus*, is the most mature specimen on public display in Southern California.

Eleven species of the fan-palm genus *Brahea* from Mexico and one from Central America grow in the garden, many of them successors to trees grown earlier from seed collected south of the border. From Baja California come the tallest brahea, San Jose hesper palm (*B. brandegeei*), and the blue hesper palm (*B. armata*). The Guadalupe Island palm (*B. edulis*) is endemic to the island of that name 160 miles off the Mexican coast. Other species range variously from Sonora and elsewhere in Mexico, while *B. salvadorensis* is from Central America.

The massive gray trunk of the Chilean wine palm (*Jubaea chilensis*) is topped with stiff featherlike foliage, dull green above and grayish beneath.

Opposite: Senegal date palms (*Phoenix reclinata*) frame a bench placed for a view of the lower part of the palm garden.

Two other fan palms grow in the desert garden. California fan palm (*Washingtonia filifera*) and the Mexican fan palm (*W. robusta*) are found in desert oases with springs and underground water. The Mexican species was widely planted in Southern California long before anyone realized it would grow so tall and out of scale on city streets that the irreverent would call it a "telephone pole with a fright wig"—useful if unkind horticultural information.

The palmettos, genus *Sabal*, are all from the Americas. Palmetto (*S. palmetto*), native from South Carolina to Florida, Cuba, and the Bahamas, has a trunk twenty to ninety feet tall, edible fruit, wood for building, and leaves for roof thatching, baskets, mats, and hats. The Rio Grande palmetto (*S. mexicana*), native from Texas to Guatemala, has a stockier trunk, and the dwarf species *S. minor*, native from Georgia to Florida and Texas, has a six-foot or shorter trunk.

The so-called saw palmetto, not a sabal but *Serenoa repens*, may derive its common name from a shared habitat with the sabals and a resemblance to *Sabal minor*. Slow-growing, often in a dense knee-high scrub, it is native from South Carolina to the Florida Keys and coastal Texas.

One of two Asian fishtail palms, *Caryota ochlandra*, from southern China, is seen in the palm garden. The other, *C. urens*, from south Asia, is in the jungle garden nearby and near the Huntington Gallery's north door. These handsome, fast-growing trees have a life span of only about twenty years, producing inflorescences or flower clusters only in their final years.

Fine examples of Senegal date palm (*Phoenix reclinata*) stand out near the entrance pavilion. Several palms contribute to the Huntington Gallery landscaping, among them the tenderest of all, an unidentified species of royal palm (*Roystonea*) from the Caribbean. Other palms are planted in the North Vista, along Mausoleum Road in the north part of the gardens, and along Euston Road, bordering the gardens on the south.

ABOVE: Tall California fan
palms (*Washingtonia
filifera*), the sole palm
native to California,
dominate the skyline of
the desert garden. The
Canary Island date palm
(*Phoenix canariensis*) in
the middle of this picture
was moved from Collis
Huntington's house in
San Francisco to San
Marino Ranch.

LEFT: Here behind
the short form of the
multitrunked *Chamaerops
humilis* is the tall form.
The Canary Island date
palm, the most widely
planted large palm in
California, is at right.

• THE DESERT GARDEN •
PLANTS FROM EXTREME ENVIRONMENTS

Spiny plants are close to the walkways in the desert garden, tempting visitors to touch at their peril. The rockery lining the slope of the main path (left) was constructed in 1927 of lava rock from Arizona and contains many species of *Mammillaria,* commonly referred to as pincushion cacti. Golden barrel cacti *(Echinocactus grusonii)* are seen in the background and in the small picture at right.

AT FIRST, PEOPLE unfamiliar with the planet's arid regions are likely to find the desert garden's plants bizarre. Then, like anyone who has been beguiled by the austere beauties of such places as Baja California, the deserts of mainland Mexico, Africa, or our own Desert Southwest, they discover the collection to be an endless delight. The garden displays, as few others anywhere do, a rich variety of plants of bold geometric and sculptural shapes, strong textures, and some of nature's most dazzling color in flower and leaf. Equally important, it reveals the astonishing ways by which desert plants cope with drought and defend themselves against enemies. The garden distinguishes Henry Huntington as a great collector for bringing together a major plant group largely unknown and little appreciated in his time, when deserts were widely thought to be wasteland of little value.

Begun as a cactus garden and later expanded to the broader category of xerophytes, or aridity-adapted plants, it grew to preeminence and today remains one of the world's finest. It has also become one of the Huntington's most botanically important gardens. Many of its specimens, grown over the years to unusual size, are now all the more valuable for science and education because curators have kept precise records of their origin and growth. Today the gar-

den and a conservatory, to be noted presently, contain more than five thousand species, half the plants in the world considered to be succulent—that is, plants that store water in leaf, stem, or root.

More than a showcase, the garden also functions as a laboratory for the study of physical properties of plants and their adaptations to growing conditions. It is also a place where new plants can be tested for landscape value, selected, hybridized, and propagated for introduction to the horticultural trade.

It can also supply material for research on desert plants for food, pharmacology, and industrial uses, continuing an age-old pursuit. As a sampling, *Aloe vera* produces a healing, cosmetic agent; the dragon tree (*Dracaena draco*) exudes a red "dragon's blood" resin used as a violin wood preservative in Italy; euphorbias yield products for such varied uses as commercial waxes and rubber and poisons for catching fish or tipping arrows; jojoba (*Simmondsia chinensis*) yields an oil for high-tech uses to replace sperm whale oil. Many plants have had multiple uses. Early peoples in Mexico used agaves, for example, for food from edible root crowns and flower stalks; as fiber for fabric, paper, thread, and cordage; and for alcoholic drinks fermented from stem sap to make pulque, in turn distilled to make tequilas and mescals. In our time agaves produce material for formulating such useful drugs as cortisone and the hormones estrogen and progesterone.

The garden is not intended to simulate desert scenes but rather to explore botanical, horticultural, and landscaping aspects of xerophytes from the world over. It began in 1905, when William Hertrich persuaded Henry Huntington to try cacti

In summer, satin yellow flowers appear amidst golden spines on the twisted, deeply grooved columns of star cactus (*Astrophytum ornatum*).

in an area of poor soil that resisted conventional planting. Though reluctant because of a previous personal encounter with cactus spines, Huntington agreed to an experiment on half an acre, and Hertrich gradually moved in some three hundred cacti. That included a mature *Cereus xanthocarpus*, old when installed and now the garden's oldest and most massive cactus. In 1908 Hertrich brought back from Arizona three carloads of saguaro (*Carnegiea gigantea*) and other cacti; many are still in place, though today young specimens replace saguaros lost in a freeze in 1937. In 1912 Hertrich brought two carloads of cacti and other succulents from Mexico. That year the garden expanded to five acres and in 1925 it acquired five more acres on fill over a former irrigation reservoir. Specimens of *Yucca filifera*, the tallest yucca, still remain from plantings around the shore, an irregular line of venerable trees through the lower garden. In 1927 Hertrich planted a bank of pincushion cacti (*Mammillaria*) beside the central path in more than a thousand linear feet of scoria, a porous red basalt, a planting destined for fame decades later as one of the world's premier rockeries. After a lull during the Great Depression and World War II, momentum resumed in the 1960s, with the garden's scope enlarged to include desert species besides cacti. In 1981 the area grew to its present twelve acres, and in 1985 a desert garden conservatory opened to the public.

The garden has sixty plant beds, numbered for identification, some of them with two or more subdivisions. It straddles the same south-facing fault scarp that benefits other botanical garden segments with a sunny slope and a cold-air flow

The central part of the desert garden displays a medley of cacti including short, round golden barrel cacti in front of the saguarolike *Trichocereus pasacana* (on the right); yellow flower spikes of the seldom-seen *Agave bracteosa* (to its right); and *Yucca filifera* (the original plants that bordered Huntington's duck pond). At left is an old treelike specimen of *Cereus alacripotanus* and a hybrid of *Cereus huntingtonianus*. The green succulent carpet below the cereus cacti is *Sedum palmeri*.

Despite their spines, some of the most formidable desert plants produce blooms of utmost delicacy.

TOP LEFT: Pink flowers of *Echinopsis* 'Stars and Stripes'

TOP RIGHT: The red and orange-tipped, tulip-shaped flowers of *Echeveria setosa* sit astride rosettes of furry leaves, which spring from succulent stalks.

BOTTOM LEFT: Seed capsules of *Cheiridopsis roodiae*, reminiscent of passion flower, open to release their cargo for germination.

BOTTOM RIGHT: The tiny yellow flowers of the crown of thorns euphorbia (*Euphorbia milii* var. *splendens*) lack petals but are encircled by showy red bracts.

Skillful planting in the lower part of the desert garden contrasts the horizontal mats of low-growing succulents—crassula, sedum, lampranthus, and echeveria—with the vertical forms of columnar cacti, tall *Yucca filifera*, palms, and at center, the orange flower spike of sotol (*Dasylirion longissimum*).

that on frosty nights can produce a critical temperature differential of five degrees between upper and lower reaches. The northern high ground is devoted mostly to the Old World, the western and southwestern parts of the garden to North America, and the low southeastern part to South America.

Most specimens have been planted where they will grow best, but a long-term effort is under way to reorganize some beds to represent regional flora. The first, Baja California, was followed by beds devoted to the Sonoran and Chihuahuan deserts of Mexico and our Southwest, California deserts, various South American regions, the Canary Islands, and Madagascar.

Let us look now at a few garden highlights, proceeding roughly from high to low and north to south.

Near the conservatory in the upper section, home to South African succulents, tower mature, treelike *Euphorbia ingens*, remnants of a once-larger stand. These are not cacti but succulent euphorbias that resemble columnar cactus species seen

In the foreground the pinkish red stems of *Puya venusta* (left), with their pink and dark purple flowers, grow high above silver-leaved rosettes. For much of the year this rare xerophytic bromeliad, *Puya chilensis* (right) seems an unpromising clump of spiky grass, but in spring it produces floral stalks covered with spectacular chartreuse flowers to lure its pollinators.

elsewhere in the garden. Nearly all the cacti are native only to the Americas, so the comparison makes a graphic example of convergent evolution, in which unrelated plants have evolved similar adaptations to similar environments in different locations. Nearby, the small milk barrel euphorbia (*E. horrida*) likewise resembles a barrel cactus. Flower size offers one way to distinguish between look-alike euphorbias and cacti: those of euphorbias are usually small and inconspicuous, those of the cacti are typically large and showy.

In the same area, the crown of thorns euphorbia (*E. milii* var. *splendens*), native to Madagascar, shares its common name with a species more likely to be the one mentioned in the Bible: the thorny shrub *Zizyphus spina-christi*, planted behind the euphorbia.

Two hundred species of aloe throughout the upper garden stage the Huntington's most exuberant color spectacle in winter with flower clusters of brilliant red, orange, and yellow. Mostly from South Africa, some aloes are stemless, some are trees, all are leaf succulent—that is, they store moisture in their leaves. Near the conservatory, several tree aloe (*Aloe bainesii*) specimens stand out; tallest of the aloes, they can reach fifty feet in their native habitat.

Near them is one of the most useful of the genus *Agave* for landscaping, the gray-green *A. attenuata*. Planted close to it is a unique blue variant, *A. attenuata* 'Nova', brought back from Mexico in 1970 by a Huntington team, that shows promise as a garden subject. The desert garden contains nearly two hundred of the three hundred known agave species. Most are mistakenly called century plants because of the long interval before they put out flower stalks, but that period is more like seven to thirty-

five years than a hundred. Spectacular and as tall as thirty feet, those stalks spring from basal leaf clusters called rosettes, which may be a few inches to fifteen feet high, depending on species. Agaves range from the southern United States to South America; most come from Mexico.

A specimen of special interest is an *Agave celsii* in the Sonoran bed near the garden's lower entrance. It is listed as No. 1 on the botanical gardens' survey and inventory, begun in 1930—a list that by 1995 had passed 80,000 entries.

Crassulas, echeverias, kalanchoes, sedums, and other members of the crassula family spread varied colors of foliage and flower across the garden. One of the most unusual is a black aeonium (*Aeonium arboreum* 'Zwartkop'), dramatic with dark purple, almost black leaves ("zwartkop" means "black head" in Dutch).

One of the desert garden's most spectacular beds contains some sixty species of pincushion cacti (*Mammillarias*) from Mexico. The two outstanding are *M. compressa,* forming mats that creep over the rocks, and *M. geminispina*, snow-white and mounding, suggesting a bubbling cascade as surely as white pebbles do in a skillfully executed Japanese dry-garden watercourse. In the same location, the golden barrel cactus (*Echinocactus grusonii*), a familiar desert plant, creates one of the garden's most dramatic landscape features for its numbers, form, and color. To cope with periods of drought it has accordionlike ribs that let the body expand as it stores water and contract as it uses it. Many of these slow-growing cacti, in place since 1925, will weigh several hundred pounds when their moisture content is high.

Nearby are two striking examples of caudiciforms, a term that refers to a water-and nutrient-storage organ in the stem or the root, at or below ground level.

The brilliant purple daisy flowers of *Lampranthus amoenus*, one of the taller-growing species in the ice plant family, open at midday in early spring. Above rises the unmistakable spike of sotol.

OPPOSITE: In the Baja California planting bed, creeping devil (*Stenocereus eruca*) grows spiny stem segments that spread across the ground to form a spectacular ever-expanding patch, intimidating to intruders.

One is the elephant-foot yucca (*Yucca elephantipes*), which takes its name from the shape of swellings low on its multiple trunks. The other is the Mexican *Beaucarnea recurvata*, also with a distended base to its fissured trunk, planted in an impressive grove along the west edge of the garden.

In the Baja California bed, the boojum tree or cirio (*Fouquieria columnaris*) is arguably Baja California's most unusual tree. Its erect, tapering trunk will bristle briefly with tiny leaves after a rain and it may sprout branches of unpredictable form at the apex. Here planted in a grove, it evokes Lewis Carroll's fantasy creature, the boojum, in his *The Hunting of the Snark*. Its other name, cirio, is from the Spanish for taper, referring to its candle shape. Cirio is related to the red-flowered ocotillo (*F. splendens*) of the Sonoran Desert, and Adam's tree or *palo adán* (*F. diguettii*) of Baja California, also seen here. The Baja California bed also has agaves, nolinas, dudleyas (*Dudleya brittonii*), and such cacti as cardon (*Pachycereus pringlei*), pitahaya agria (*Stenocereus gummosus*), and the remarkable creeping devil (*S. eruca*), a prostrate species that produces extensive thickets.

Not far away a large specimen *Nolina matapensis*, from the Mexican state of Sonora, with a crown of thick, grasslike leaves, is rare in cultivation. It shares a bed with yuccas and several cacti. One of these, *Opuntia ellisiana* from Mexico, has flattened padlike stems on which the scale insect cochineal feeds; when dried and ground the insects produce a red fabric dye. Stem-succulent like barrel cacti, prickly-pear opuntias have padlike stems that function as leaves. Fruits of opuntias are edible and in some species the pad, the *nopal* of Mexico, is also eaten cooked as a vegetable.

The California desert bed illuminates the meaning of "exotic" as a plant that is native somewhere else. The Joshua tree (*Yucca brevifolia*), native to the Mojave Desert

LEFT: The waxy orange flowers of prickly pear cactus (*Opuntia littoralis*) form on the circumference of their flat, spiny padlike stems.

RIGHT: The blue leaves of *Kalanchoe grandiflora* are arranged in an intricate geometric pattern.

OPPOSITE TOP: Among clustered red and green jelly-bean leaves of Christmas cheer (*Sedum ×rubrotinctum*), a flourish of yellow star-shaped flowers appears in winter.

OPPOSITE BOTTOM: Thorns paired along the margins of *Euphorbia grandicornis* make the fleshy stems impenetrable.

thirty miles north across the mountains, is exotic in San Marino and does not fare well here. On the other hand, Our Lord's Candle (*Y. whipplei*), native to the San Gabriels and coastal mountains, does do well, regularly appearing as seedlings in the garden.

Likewise exotic here are plants from the low desert a hundred miles away. An example of a tree from the desert but not a xerophyte is the California fan palm (*Washingtonia filifera*). It grows in oases in California, western Arizona, and northern Mexico that have water from springs, an underground stream, or a high water table. The related Mexican fan palm (*W. robusta*) occupies similar habitat from the border to much farther south. Other plants in the California bed include brittlebush (*Encelia farinosa*), creosote bush (*Larrea tridentata*), and beavertail cactus (*Opuntia basilaris*).

Planted in several lower-garden beds, two rare xerophytic bromeliads, related to the pineapple, come from Chile. They bloom in the spring, *Puya alpestris* with flowers of ethereal blue-green, and *P. chilensis* with vivid chartreuse flowers. Both have hooked spined leaves with sawlike edges. Beside the main path in the middle of the garden another member of the pineapple family, *Abromeitiella brevifolia,* is seen clambering over the rocks in a giant mound.

Finally, succulent ground covers are planted in four beds as a border to the road along the desert garden's west side. Here cotyledons, crassulas, echeverias, senecios, and sedums provide year-round color. Mesembryanthemums, or midday-blooming flowers, including lampranthus, explode in March with some of the most intense of all flower colors.

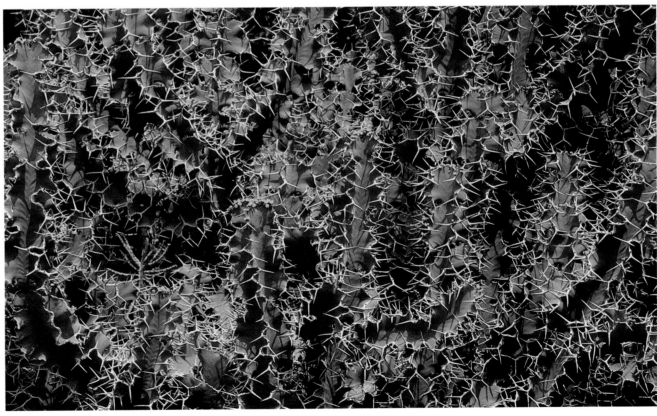

TOP LEFT: Creamy flower clusters of *Yucca filifera* dangle from angled branches with a top-knot of bladelike leaves.

TOP RIGHT: The young leaves of *Agave nelsonii* retain the imprint of their nascent thorns.

LOWER LEFT: The distinctive red-orange and yellow poker-shaped flowers of *Aloe africana* are borne on short stems.

LOWER RIGHT: Part of the garden's showiest color is the winter flowering of two hundred species of aloe. Various hybrids are growing beneath a towering dragon tree.

OPPOSITE: Aloes range in size from the small single rosettes of *Aloe humilis* to *Aloe* 'Rooikappie', seen here in the foreground, to the gracefully branched tree aloe (*A. bainesii*), tallest of the aloes. The succulent swordlike leaves of aloes are similar to those of agaves, but they do not die down after the plant has flowered.

The Desert Garden Conservatory
Shelter for the Fragile and Delicate

Welwitschia mirabilis, or marvelous welwitschia, has only two leaves, but they may live for more than a thousand years.

THE DESERT GARDEN is limited to plants that can grow in the San Gabriel Valley climate. The Desert Garden Conservatory, a greenhouse at the upper end of the garden, broadens the possible range of plants in the collection by providing the kind of shelter perhaps a third of the world's succulent plants need to thrive here or even survive. Opened in 1985 for public viewing, it holds some 3,000 specimens under controlled conditions of temperature and irrigation and protected from trampling as well as competition by more vigorous plants.

Its specimens are grown in special unglazed stoneware pots in a potting mix designed for drainage and moisture retention, with a top dressing to suppress weed and algal growth. Keeping out birds and insects protects plants from unwanted hybridization; so does hand pollination, done with small paint brushes. One conservatory activity, propagating and distributing rare species, reduces demand among collectors for plants from the wild, helping to preserve them in nature.

The conservatory's display is laid out on seven benches and a landscaped bed at the west end, while containers for large specimens are placed on the floor near the center. More than thirty-five plant families represented in the collection include the seven major families, which comprise seventy percent of all succulent species, and some of the more obscure. New World plants are grouped mostly south of the center aisle and Old World plants—from Europe, Africa, and Asia—are north of it and in the plant bed. Cacti are mostly on the south side, because they are native to the New World. The exception, *Rhipsalis baccifera,* is displayed along with other epiphytic cacti on the north side.

Many are the curiosities to be seen here. Among them are plants that grow almost entirely underground except for their leaf tips, getting light to the interior with so-called windows in the tips. The lithops, "living stones," bear no resemblance to any other kind of plant; all that is visible are fleshy leaves that mimic the shape and color of rocks or soil in their native habitat. Another mimicry adaptation

Some of the thirty-five plant families represented in the conservatory are shown in the picture above. Specimens are grown in stoneware pots in special potting mix.

RIGHT: The jewel flower of Mexico (*Tacitus bellus*) was discovered and introduced into cultivation in the early 1970s. It is part of the Huntington's extensive research collection of New World Crassulaceae, the crassula or stonecrop family.

comes from the *Ariocarpus*, cacti that avoid being eaten by resembling jagged stones. Perhaps the most remarkable of all the plants here is *Welwitschia mirabilis*, from the rainless desert of Namibia and Angola. It can live for more than a thousand years with only its original two leaves, which eventually grow to ten feet long on a trunk several feet wide but less than a foot high. A distant relative of the conifers and cycads, it is a so-called living fossil, producing seed from cones, not flowers. The largest plants here, started from seed in 1970, produced cones and seed only after fourteen years.

The planting bed at the west end features large Old World succulents. Prominent here are ten of the eleven species of Didiereaceae, a small family of succulent shrubs endemic to Madagascar, with sprawling, thorny stems that resemble those of the unrelated native of southwestern U.S. deserts, the ocotillo.

Opposite is a sampling of stapeliad flowers, succulents in the milkweed family: *Caralluma rogersii,* top left, with delicate glasslike hairs in the flower and an unpleasant scent; *Stapelia flavopurpurea,* top right, with sweetly scented flowers; *Orbea semota* var. *lutea,* bottom right, with brilliant yellow flowers and delicate hairs that flicker in the faintest breeze, perhaps as an attraction to pollinators; and *Huernia nouhuysii,* bottom left, with no fragance at all.

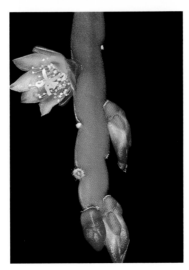

Lithops, this page top, the "living rocks", put out flowers from fissures across the middle. Many small, globular stemmed species of cacti, center row, make ideal container specimens: *Notocactus graessneri* var. *albisetus,* left, of Brazil is unusual for its bright green flowers. *Borzicactus weberbaueri,* center, of Peru has showy tubular flowers pollinated by hummingbirds. *Mammillaria pilcayensis,* right, is one of a genus of more than 150 species collectively known as pincushion cacti. *Rhipsalis boliviana,* bottom, collected on a Huntington expedition to Bolivia in 1984, is part of the Huntington's extensive research collection of epiphytic cacti.

IN THE GARDENS, PART II

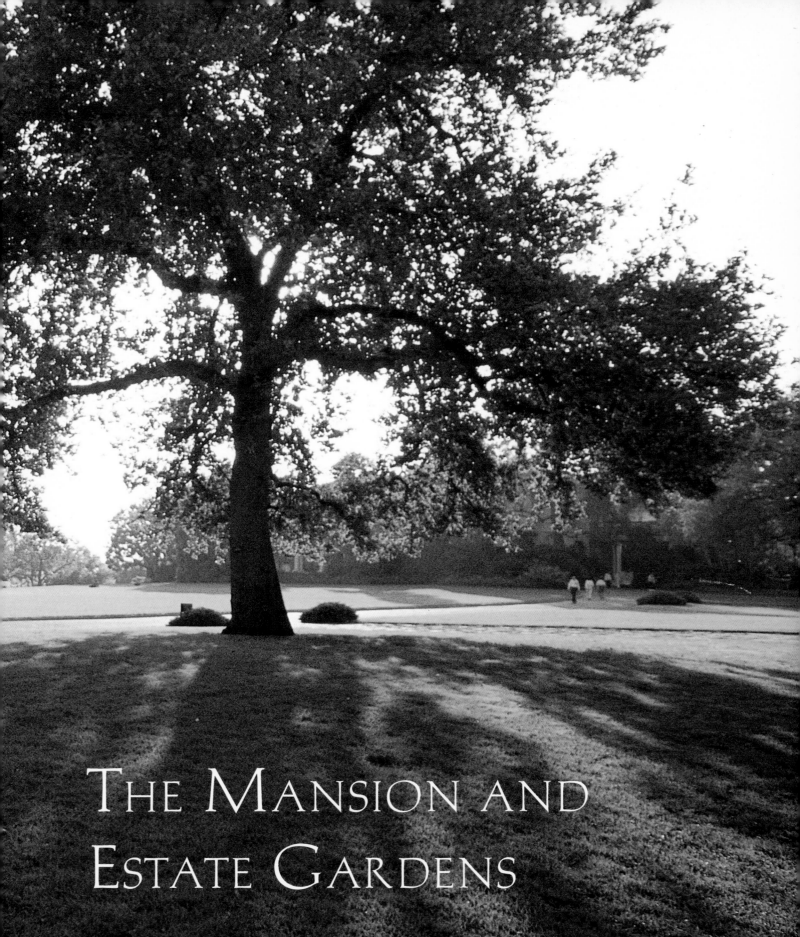

THE MANSION AND ESTATE GARDENS

THE HUNTINGTON GALLERY GARDEN
LANDSCAPING AROUND A MANSION

This plant screen on the north and east side of the loggia was intended to create a partial enclosure so the space could be used as an outdoor room. *Philodendron bipinnatifidum* and Senegal date palm (*Phoenix reclinata*) have grown large and luxuriant.

PAGES 71–72: Late afternoon shadows on the Library lawn

THE MANSION, now the Huntington Art Gallery, occupies the heart of the most intensively developed portion of the gardens. Seen from its south terrace, the property's grand estate scale is still apparent in expansive views across lawns on three sides. To the east, the prospect reaches to the entrance and palm gardens. A broad crescent formed by tall-tree boundaries of the jungle and subtropical gardens extends southeast to southwest as far as the rose garden west of the house. North of the house the landscape elements are closer together: a circle formed by a porte-cochere entry road, the stately North Vista and its flanking camellia gardens, and the Shakespeare garden.

As founder Huntington intended, the gardens—even when large-scale and formal—retain a so-called softscape character, that is, plants are more important in landscaping than structure. Plants create the setting for the mansion, rather than such architectural extensions proposed by architect Myron Hunt, but not accepted by Huntington, as balustraded terraces down the slope from the south terrace. Even the North Vista's formal elements are

In the view from the mansion's south terrace, a long line of mature trees in the subtropical and jungle gardens screens off the intensively urbanized San Gabriel Valley. The terrace surveyed a rural scene in the first decades of this century when the house was new.

OPPOSITE: The pure white trumpet flowers and spotted leaves of calla lilies (*Zantedeschia aethiopica*) surround a stone urn with bas-relief figures near the Huntington Gallery entrance.

executed with lawn rather than paving and palms rather than structure, used architecturally at the sides. Planting defines the larger themed gardens all around, so they flow together mostly without such intervening works as walls, fences, and gates—though they are used in the Japanese and two smaller gardens.

The estate house has a door and a porte-cochere entry on the north side. Today the public entrance is at the east end through a loggia that the Huntingtons used as an outdoor room. Dense with large plants, the landscaping concentrates around these northeast-corner features. Standing out among trees are spreading coast live oaks (*Quercus agrifolia*) in the circle, Senegal date (*Phoenix reclinata*) and rhapis palms, a tall white silk floss tree (*Chorisia insignis*) with a swollen green prickle-studded trunk, red-flowering erythrinas or coral trees, and a huge white-flowering *Talauma hodgsonii,* a relative of the magnolia from the Himalayas. In the planting's lower story, bird of paradise (*Strelitzia reginae*) and giant bird of paradise (*S. nicolai*) are prominent, but the stars of the show are cycads (see pp. 82–85). These primitive palmlike plants have grown into a thick screen in a rockery north of the loggia and are planted in several other locations. Here also are Australian tree fern and several other fern species. South of the loggia, a hundred-year-old English

The tufted heads of cycads (*Cycas revoluta*) and a majestic Moreton Bay chestnut (*Castanospermum australe*) flank a passage from the balustraded terrace to the broad lawns. The tree's orange and yellow flowers attract birds and butterflies in summer.

yew (*Taxus baccata*) forms a great green mass.

Running the length of the house, the broad south terrace has a balustrade topped by a line of ornamental urns. It also has clipped round pillars of Australian brush cherry (*Syzygium paniculatum*) that repeat the façade's architectural elements. Rearing south of the balustrade, two giant coral trees (*Erythrina caffra*) partly shade the terrace, as does a Moreton Bay chestnut (*Castanospermum australe*) at the southeast corner. At the east end, steps lead down to a lawn and planting that features staghorn ferns (*Platycerium*) clinging to the terrace retaining wall.

West beyond the terrace in a protected corner of the house, a royal palm (*Roystonea*) of unidentified species from the Caribbean flourishes as yet unharmed by cold wind or frost. Other areas around the mansion—the rose garden on the west, and the Shakespeare garden and North Vista on the north—are discussed in the next three chapters.

Ferns, bird of paradise, and cycads thrive in this shady rockery next to the mansion's loggia. Hertrich intended the rockery for plants requiring semishade and protection against frosts. He found the porous rock, a soft tufa stone, near Santa Cruz, California. The Italian marble fountain is topped by a bronze statuette, *Bacchante* by Frederick MacMonnies (American, 1863–1937).

The large silhouette of a philodendron leaf in the foreground and the distinctive shape of cycads in the border animate this view of the mansion's south terrace.

Plantings at the northeast corner
of the Huntington Gallery contain
a thick understory of cycads,
ferns, and other plants in the
shade of exotic trees.

Opposite: Seen from above, the south
terrace is a so-called hardscape, an
outdoor space defined by structure
and geometry; even such shrubs as
Australian brush cherry (*Syzygium
paniculatum*) are clipped as pillars.
Compare this with the softscape
North Vista, pp. 86–90, a formal space
rendered primarily in plant material.

Intervals of green lawn serve as graceful transitions from one specialized, richly planted garden to another. This one separates the rose garden (displaying a profusion of the Floribunda rose 'Showbiz') and the varied skyline of the subtropical garden beyond.

OPPOSITE: On this sunny November day a few pink blossoms of the early-flowering *Camellia japonica* 'King Size' can be seen at left. A cycad juts out just beyond the statue, inviting comparison with the similar-looking palms beyond.

• THE CYCAD COLLECTION •
SURVIVORS FROM THE ERA OF DINOSAURS

ZOOLOGISTS CALL the Mesozoic Era the age of dinosaurs. Botanists call it the age of cycads, plant-world survivors known from the fossil record to have outlasted dinosaurs by 65 million years since those creatures met extinction. Cycads are the most primitive family of seed-producing plants, intermediate in evolutionary development between ferns and conifers, to which they are related. Like conifers, they do not produce flowers but bear their seed in cones. The Huntington collection has representatives of nine of the eleven cycad genera and half of the 150 or so known species. They appear in the landscaping of the Huntington Gallery and the Japanese garden as well as in the Australian, jungle, and desert gardens.

William Hertrich began the collection in 1910 with a few plants of *Ceratozamia mexicana* and *Dioon edule* that he brought back from Mexico. In 1913 he received a shipment of exceptional specimens, some reputed to be three to four hundred years old, from Japan. That same year he made a major purchase from collector Louis Bradbury of nearby Duarte. Later, the widow of Los Angeles oil baron Edward L. Doheny contributed a specimen of the exceedingly rare *Dioon merolae* to the Huntington collection.

Young cycads start out with the airy, lacy appearance of ferns then eventually age into something resembling palms. Indeed, the most commonly grown cycad in Southern California, *Cycas revoluta*, is misleadingly called sago palm. The taller woody cycads grow very slowly, adding a new rosette of leaves at the top of a thick, columnar trunk each year, or none in cone-bearing years, so a plant might take more than a century to reach its ultimate ten-foot height. Pollen-producing male cones and seed-bearing female cones appear on separate plants, though not necessarily at the same time, so even with a male plant nearby a female may remain infertile.

Cycads are largely found in limited areas of Australia, Japan, Southeast Asia, South Africa, and the Americas. Some grow in shaded tropical forests, others grow in dry, open country where their leathery leaves are a response to arid conditions. Such genera as

Cycads are palmlike in appearance and in name ("cycad" derives from the Greek "cyckos," which means palmlike) but are more closely related to conifers. What appear to be cycad flowers are actually cones, which often have intricate shapes.

Cycas are widely distributed throughout Southeast Asia and northern Australia. More restricted in distribution are such others as *Encephalartos* and *Stangeria,* found only in South Africa, and *Bowenia, Lepidozamia,* and *Macrozamia,* found only in Australia.

Their form and beauty have made cycads so popular as ornamentals and collector's items that to supply the market many of them have become rare and endangered in their native habitats. In South Africa, for example, there may be more cycads in private gardens than are left in the wild. Some measures have been undertaken to protect natural populations. In South Africa, people are now required to have a permit to own cycads. In Australia, where for years ranchers destroyed macrozamias because of their toxicity to livestock, a Macrozamia National Park provides protection within its boundaries.

Cycad cones can sometimes resemble pineapples and weigh as much as eighty pounds. The smaller cones of *Encephalartos* (above) suggest pine cones. Male and female cones are produced on separate plants. *Cycas revoluta* (left and opposite page) has sago starch in its stems, giving the plant its common name of sago palm.

• THE NORTH VISTA •
REINTERPRETING A FORMAL TRADITION

Stone dolphin detail on the baroque fountain

GRANDEST of the Huntington's gardens though little more than an acre in size, this grassy allée reaches from the Huntington Gallery north six hundred feet—the length of two football fields—to focus on an early baroque fountain backed by greenery and a spectacular wall of mountains. In a long European tradition of axial outdoor spaces related to buildings, it was the conception of Myron Hunt, architect of the Beaux Arts-style manor, who liked to design gardens for his estate houses. Nonetheless this space was defined not by the structures expected in formal gardens of that time but by planting.

Several key elements establish the spirit of a garden from the baroque period, 1600 to 1750. One is the Italian stone fountain, complete with sculptured dolphins, that Henry Huntington bought knocked-down in England and had shipped to San Marino for reassembly in a basin thirty-eight feet across. Another is a straight row of graceful fountain palms (*Livistona australis*) on either side of the seventy-five-foot-wide lawn. Arrayed architecturally against native oaks, their trunks form a living colonnade (with some gaps on the east side where a few of the original eleven palms are gone) that directs the eye toward the fountain. Sometime after palms were planted, eleven eighteenth-century Italian stone figures were placed

Twenty-two Italian garden statues were each
matched to a stately palm on either side of the
rectangular part of the North Vista's long
lawn—though some palms on the east side
were later lost—and others were placed around
the wider lawn at its south end, shown here in
the foreground.

North Vista 1907

North Vista 1920

Native oaks dominated the site that was to become the North Vista (left). Years later palms surrounded by groomed hedges lined the allée (right) and a splendid baroque fountain (opposite) had been installed as the focal point at the north end.

on each side of the straight greensward, inward of the palms and repeating their rhythm. Later still, camellias filled in between palms and azaleas between statues, to create a flowering base for the colonnade. Two stone figure pairs, probably mid eighteenth-century German, mark the grass rectangle's south end near the mansion, and disposed around an irregular lawn south of that are six other single figures and an Italian tempietto holding a nineteenth-century sculpture.

Through the oak woodland on three sides of the North Vista, pathways wind among plantings of camellias (see pp. 92–97), azaleas, and other shade-loving plants. Evergreen azaleas along the allée and the paths emphasize Southern Indicas. On the west side are sports of the Belgian Indica 'California Sunset'. Near the north entrance to the east walkway are Satsuki azaleas, late-blooming plants that adapt well to bonsai culture. At the southeastern edge are several locally developed azalea cultivars that do well in Southern California. Begonias, fuchsias, gardenias, impatiens, spiraeas, violets, and several ferns are also part of the shaded understory.

THE CAMELLIA COLLECTION
A PLANTSMAN'S LIFE WORK LIVES ON

The dark leaves and abundant soft pink peony-form flowers of *Camellia hiemalis* 'Showa Supreme' (opposite), all but obscure the fountain topped by a winged cherub. *C. hiemalis,* believed to be a hybrid between *C. sasanqua* and *C. japonica,* bloom longer and later than *C. sasanqua.* They are not found in the wild.

Camellias, like roses, have been cultivated for centuries, and like roses they enjoyed accelerating popularity in the nineteenth century leading to a kind of breeding explosion in the twentieth. At the Huntington, camellias line roads and paths in some twelve acres in the central and western parts of the gardens. They are planted in native oak woodland on both sides of the North Vista and under mixed pine and oak on the slopes of North Canyon, north of the Japanese garden. With thirty-five species and twelve hundred cultivars of camellias, and others continually being added, this is one of the most comprehensive collections anywhere.

William Hertrich's lifelong work made the collection notable. Only one specimen remains from before he went to work for Henry Huntington in 1904, a splendid 'Pink Perfection' beside a path west of the North Vista, the oldest camellia on the grounds. All the rest date from Hertrich's time or later. He began growing camellias from seed in 1912 as understock for grafting new varieties and for setting out along the North Vista and in North Canyon. Forty-two years later he published the results of his long-term opportunity to propagate, study, and photograph a thousand cultivars in his three-volume *Camellias in the Huntington Gardens.*

From his thirty-year effort Hertrich had many plants available for grafting by 1942, when the Southern California

C. japonica 'Henry E. Huntington' is a cultivar developed by the Nuccio Nurseries of Altadena, and released in 1995, the Huntington's seventy-fifth anniversary as an institution.

OPPOSITE: *Camellia japonica rosea superba* has large, deep pink flowers of the formal double type. Growing usually to ten feet or so, in California one hundred-year-old plants of *C. japonica* have achieved double that height. This is still the best-known species, but its popularity is increasingly challenged by the reticulata and sasanqua camellias.

Camellia Society began to participate in the collection. After World War II the society agreed to donate plants and create a test garden, and the work of developing new cultivars began in earnest. In a close relationship with the Huntington ever since, the society has invested many volunteer hours helping maintain the camellia gardens and develop them through such activities as seed collection and an annual camellia show.

Camellias originated in Asia and after a long history of cultivation the first *Camellia japonica* reached the Western world in the early eighteenth century. Two camellias in the gardens represent the introduction of this species to England in 1792: *C. japonica* 'Alba Plena' and 'Variegata', both found along the upper walk in North Canyon. 'Alba Plena' is also found east of the North Vista. *C. japonica* has proven to be the most popular camellia species, with perhaps 25,000 named cultivars. It forms handsome shrubs or trees with broad, rounded, shiny foliage and single-to-double flowers that are seldom fragrant and range in color from white to red. Peak blooming season is January to March. One Huntington introduction is 'Margarete Hertrich' propagated by William Hertrich in 1944 and named after his wife. A medium white formal double with many small petals, it remains one of the favorite white japonicas of camellia hobbyists. It grows along the lower pathway in North Canyon. The Elegans variant cultivars of *C. japonica* give their name to Elegans Lane west of the North Vista.

Related to *C. japonica* are the Higo camellias in North Canyon. Much admired in Japan, where they are often used as bonsai plants, Higo is considered a style of flower rather than a species. They have been crossed and selected for centuries from *C. japonica* and probably the lesser-known *C. rusticana*, the snow camellia. With rare exceptions, the petals number between five and nine, may be uneven in length, and are often thick and leathery. Their most obvious features are the numerous stamens, with a count varying between 100 and 250. Sometimes matching the flower's coloring, they may appear densely packed in a central column or flattened out into a wide ring or crown, leaving the single pistil alone in the center.

A large selection of another species, *C. sasanqua,* introduced into England in 1811, is planted on the east hillside of North Canyon. Sasanquas bear profuse, small, often fragrant flowers in varied forms in white, pinks, and reds, from October into February. The plants have a more gracefully drooping, weeping, or spreading habit than japonicas. Two unusual ones are *C. sasanqua* 'Hugh Evans', with medium-pink single flowers and 'Mine-No-Yuki', blanketed with peony-shaped flowers in early fall.

In the wild, camellia species have flowers composed of a single row of petals. Over the years so many forms of camellia flowers have been selected from the wild and through hybridization that flower forms are classified in six types based on the number and arrangement of the petals.

LEFT: *Camellia japonica* 'Pink Perfection' is a formal double type with petals that overlap and hide the stamens.

TOP RIGHT: The deep rose pink 'John Anson Ford' is a semi-double, defined as having two or more rows of petals around a prominent stamen display.

BOTTOM RIGHT: 'Elena Nobile', a rose form of camellia has layered petals that reveal the stamens in a concave center when fully opened.

North Canyon and a knoll west of the North Vista fountain feature the first cultivars to be imported to North America of another important species, *C. reticulata*. Reticulatas are noted for huge pink, scarlet, and sometimes white, mostly semi-double flowers, peaking March through May. Some specimens are the original plants imported in the mid-1940s from Kunming Botanical Garden in Yunnan Province, in mountainous south China, where reticulatas have been cultivated since A.D. 900.

C. reticulata first appeared in England in 1820 when Captain Richard Rawes, an East India merchantman, arrived with a number of camellia plants, among them a botanically unknown specimen from China. This 'Captain Rawes' cultivar first flowered in 1826, creating a sensation with its very large semi-double rose pink flower. It can be seen along a path west of the North Vista. While reticulatas produce some of the most spectacular flowers in the genus, they grow as loosely branched, lanky shrubs or trees (up to fifty feet in height), less compact and attractive in habit than *C. japonica*.

Seven specimens of China Tea (*C. sinensis*), origin of common green and black beverage teas, are located next to the waterfall in North Canyon, so visitors can compare tea with the garden camellia. When taxonomists established the Tea fam-

TOP LEFT: *Camellia reticulata* 'Otto Hoppler', a semi-double.

TOP RIGHT: 'Wilbur Foss', a full peony form with a domed mass of mixed irregular petals, petaloids, and stamens.

BOTTOM LEFT: 'Berenice Boddy', a single form, which is characterized by one row of not more than eight regular, irregular, or loose petals and conspicuous stamens.

BOTTOM RIGHT: *C. elegans* 'Splendor' is an anemone form, which usually has one or more rows of large outer petals lying flat or undulating and a convex mass of intermingled petaloids and stamens at its center.

ily, they placed *Camellia* as a related member of *Thea sinensis*. On further analysis, many taxonomists came to believe that the tea plant shared so many characteristics with other species in the genus *Camellia* that today it is grouped with the camellias.

Other species growing near the waterfall include *C. chrysantha*, a rare yellow camellia, and *C. lutchuensis,* the most fragrant species, across the pond. *C. saluenensis,* important for developing new hybrids, is seen on the North Vista's east walk, and *C. granthamiana,* with large white flowers, at the southeast end of the North Vista.

The Huntington has made some notable introductions. One is 'William Hertrich', a large, deep cherry red, semi-double seedling of a reticulata imported from China. Other well-known introductions include 'Howard Asper', with salmon pink flowers; 'Flower Girl', with bright pink flowers; 'Beverly Baylies', a *C. saluenesis* hybrid with single pink flowers that bloom early and profusely; 'Betty's Beauty', a sport of the *C. japonica* 'Betty Sheffield'; 'Little Michael', a miniature, anemone flower type of *C. japonica;* 'Carl Tourje' a hybrid of *C. pitardii* ×*reticulata* with large, semi-double, wavy pink petals.

THE SHAKESPEARE GARDEN
VIGNETTE OF AN ENGLISH COUNTRYSIDE

"THE SUMMER'S FLOWER is to the summer sweet," William Shakespeare wrote in Sonnet XCIV. He might have had in mind the spirit of this one-acre garden between the Huntington and Virginia Steele Scott art galleries. Here in the gardens' main showcase of annuals and flowering perennials a bust of the Bard of Avon contemplates some of the plants that figure in his own writings and in English gardens from the late sixteenth century to our time. In that tradition the flowers bloom most profusely through spring and summer.

This is the third version of a Shakespeare garden at the Huntington. Opened in 1959 as a living counterpart to Elizabethan literary works in the Library, the first was devoted only to plants specifically mentioned in Shakespeare. Since many of those failed in the decidedly non-English climate of the San Gabriel Valley, the garden was revised in 1972 to be more attractive year-round by adding other ornamental plants in use during the Elizabethan era in England and elsewhere.

In 1984 the setting changed with the construction of the Scott Gallery on the north, so the landscape was again redesigned, this time as a naturalistic English countryside setting to serve as a foreground to that building. Graded berms and a central dell or bog—dry except during the rainy season—with a bridge, lawn, and planting beds suggest a stylized pastoral glade against a backdrop of neighboring trees. On days of heavy visitation the paths efficiently move visitors between the galleries and from garden to garden, and this space functions as an unusually agreeable central plaza.

Violet trumpet vine (*Clytostoma callistegioides*)

OPPOSITE: A sculptured bust of its patron poet presides over the Shakespeare garden and an audience of spring-blooming iris. Among the most successful bulblike plants for Southern California, iris do not require winter chilling and tolerate summer heat.

The waxy red flowers of the pomegranate (*Punica granatum*) overhang a bed planted with pink and white foxglove and mounds of yellow coreopsis.

OPPOSITE: The classical façade of the Virginia Steele Scott Gallery, framed by birch trees, is the background for the Shakespeare garden. A coast live oak provides shade for a colorful impatiens border and a planting of orange daylilies.

Small signs here and there in the beds supply Shakespeare quotations related to flowers, holly, and other plants. The garden is lush with perennials interwoven in an informal landscape with such flowers as buttercup, campanula, columbine, foxglove, geranium, goldenrod, iris, wormwood, and yarrow. Bulbs and such annuals as marigold make seasonal appearances. In the center, tall yellow spikes of verbascum (mullein) contrast with fountaining rushes and grasses and bold leaf forms. Near the south entrance, mounding feverfew (a chrysanthemum that bears small white flowers) reseeds among spreading rosemary. Though not usually thought of as a flower garden subject, an occasional fennel is allowed to flower and produce seed, and other volunteer plants often lend an element of whimsy to the scene.

Hedges and individual plants of holly, laurel, and viburnum provide masses of greenery, and flowering and fruiting trees—crab apple, lilac, loquat, peach, pomegranate, and quince—flaunt spring color. Arbors with domes that echo the central dome of the Scott Gallery mark the east and south entrances and in season support grapevines, sweet peas, or other climbing plants. An olive tree in place since last century, one of the oldest non-natives on the grounds, recalls Shakespeare's "olives of endless age" metaphor for peace and longevity in Sonnet CVII, proclaiming the writer will live on in his work.

Old-fashioned daylilies highlight the historical collection of more than 2,000 cultivars of *Hemerocallis* maintained by the Huntington. Empty here, the walks bustle with visitors most weekends in this crossroads to other gardens.

RIGHT: Mexican bush sage *(Salvia leucantha)* is a riotous mass of purple spikes and the leaves of the deciduous pomegranate *(Punica granatum)* wear autumn's golden color.

OPPOSITE: Among the annuals and perennials mingling in this summer view are yellow yarrow, crimson lychnis, and pink and white common foxglove.

THE ROSE GARDEN
A LIVING LIBRARY OF FLOWER HISTORY

WEST OF THE Huntington Gallery, the three-acre rose garden is one of the Huntington's themed units that interpret plants in terms of culture and history—along with the Shakespeare, herb, and Japanese gardens. It is also one of the three gardens most popular with visitors, along with the Japanese and desert gardens. Its blooming peaks in April, with most roses flowering sometime from March through December. At pruning time in January, visitors see how removing a third to half of the previous year's growth will channel a plant's energy in the year ahead into flower production rather than growth.

While some kinds of plants in the botanical gardens have hardly changed in millions of years, few have altered more than roses, the most widely planted shrub and one of the most varied in breeding. In this garden roses in beds and climbing on pergolas, arches, and a dome reveal a complex history that ranges over two and

a half millennia from Grecian antiquity to the present. The garden displays nearly eighteen hundred species and cultivars. They include examples from every important group in the rose's development and every Northern Hemisphere region where roses are native.

The garden was planned by Myron Hunt, architect of the estate house, as a formal space

The rose garden was originally planned as a spectacular massed display. Today the garden's forty beds have been planted with a living history of the rose. This pergola near the tea room supports climbers in shades of red, pink, and white. In the foreground is a bed filled with the yellow Floribunda rose 'Sun Flair'.

RIGHT: Named in honor of the seventy-fifth anniversary of the institution's founding, the 'Huntington's Hero' rose was propagated from a sport discovered on one of sixty bushes of 'Hero' in the rose garden's David Austin rose planting.

'Altissimo', a Floribunda shrub rose that is grown most successfully as a climber

with an east-west axis meant to match that of the mansion's central hall. When the foundation was begun in 1908, however, Henry Huntington had the axis shifted a few degrees to preserve two old oaks (later taken out). William Hertrich had already planted the roses according to the original plan, so the two axial lines never coincided. Twice the size of the one today, that first rose garden was reduced in area in 1922. At that time many flower beds were replaced by lawn and an eighteenth-century French stone tempietto with a sculpture was installed. Altered again several times over the years, the garden was last refurbished beginning in 1988. Meanwhile, after nine decades some of its specimen trees remain, grown now to monumental size. Among them are three Montezuma cypress (*Taxodium mucronatum*) from Mexico, a Queensland kauri (*Agathis robusta*), a pecan (*Carya illinoensis*), and several magnolias.

Today the garden contains forty beds planted to roses, including the climbing kinds. A walk through it reveals rose history from ancient to modern.

Two pergolas shade a walk that runs along the north part of the garden from its northeast entrance to near the Japanese garden on the west. Plants in the bed south of the eastern pergola are from the first period of rose history, from as early as a fourth-century B.C. mention by Greek historian Herodotus to the end of the eighteenth century. Typically these old roses—Alba, Centifolia, Damask, Gallica, and Moss—bloom only in spring, in flat or cupped blossoms ranging in color from white, pink, and pale red through purple. Yellow was not unknown, but a true yellow did not appear until the twentieth century. One of the most celebrated old roses was Centifolia, the "hundred-leaved rose" noted in Latin literature. The Huntington's Centifolias were developed by sixteenth- and seventeenth-century Dutch hybridizers.

Among these old roses are some mentioned by Shakespeare and in Renaissance herbals and rose books collected in the Library. Prominent in the display are the red rose of Lancaster (*Rosa gallica* 'Officinalis') and the white rose of York (*Rosa alba*). How they became emblems of opposing factions in the 1455–1485 Wars of the Roses is dramatized by Shakespeare in *King Henry VI, Part 1*. They are joined by the pink and white summer Damask rose 'York and Lancaster', symbol of reconciliation after the victorious Lancastrian Henry Tudor became king and married Elizabeth of York.

In the mild climate of Southern California, roses flower from March through December, but the garden is in full bloom in mid-April. The porticoed entrance to the tea room is scarcely visible behind a cascade of climbing roses, including 'Cécile Brunner', along the length of the pergola. Some of the beds here are planted in the traditional cottage garden way, combining roses with herbs, annuals, and perennials. The roses in the foreground include the yellow Floribunda 'Mountbatten' and the dark pink Floribunda 'Matangi'.

Two beds north of that pergola contain roses from the second major period of rose history, the nineteenth century, when China and Tea roses were introduced into Europe from Asia. China roses are usually short plants with small to medium-sized flowers, often in clusters. An early one here is 'Slater's Crimson China', the first true bright red and parent of all modern red roses. Tea roses tend to be tall, with large flowers that often droop on weak stems. Though both are tender in cold climates, both share the important ability to bloom more than once in a season. It took workers from 1800 to the 1840s before the first hardy remontant, or repeat-flowering, roses were at last hybridized from China and Tea roses. In time some two thousand varieties resulted from that revolutionary feat, among them Bourbon, Noisette, Hybrid Perpetual, Portland, and the first Hybrid Tea, 'La France'.

The third or modern period in rose history had its roots in three nineteenth-century advances: the appearance of the Hybrid Tea and Polyanthas varieties and of a true yellow rose color. Introduced in 1867, Hybrid Teas, considered to be backcrosses between Hybrid Perpetual and Tea roses, were destined to become the twentieth century's most popular roses, with thousands of varieties known. Introduced in France around 1870, Polyanthas were hybrids of *Rosa multiflora* by a China rose and became popular as short, compact plants with large clusters of small flowers. In the 1880s, a French cultivar, 'Soleil d'Or', crossed with Hybrid Teas became the parent of all modern yellow and bicolor roses.

Most of the garden's beds are devoted to Hybrid Tea, Polyantha, Floribunda, Grandiflora, Miniature, and other modern roses. Floribundas, pioneered in Denmark in the 1930s, are descendants of the Polyanthas with small stature and cluster

LEFT TO RIGHT: The climbing Floribunda 'Winifred Coulter Climbing' threads through one of the specially designed metal arches. A sport of the shrub rose 'Winifred Coulter', it flowers continuously and grows as high as fifteen feet. 'Cherry Vanilla Grandiflora' is pink with a golden heart. The apricot-colored 'Norwich Castle' is a relative newcomer, raised in 1979 by Peter Beales in Britain.

OPPOSITE: The velvet-red climbing Tea rose 'Noëlla Nabonnand'

bloom, used here massed as low borders to front taller varieties. Grandifloras are a cross of Hybrid Tea with Floribunda.

Beds on both sides of the walk south from the center, at the restaurant entrance between pergolas, and west of the tempietto contain Floribunda and Hybrid Tea or Floribunda alone. West of that walk are eight beds of Hybrid Teas, Floribundas, and Grandifloras introduced since 1950.

East of the walk are two beds planted to very different roses. One of these contains roses from the California gold country planted during and after the 1849 Gold Rush and overlooked until our time, when Clair Martin, curator of the rose collection, found them growing in old gardens, cemeteries, and other locations. Most are no longer known commercially, but have been identified from early rose books and catalogs in the Library. The other contains English roses, a class of large shrubs developed since 1961 by English hybridizer David Austin. He combined the form, fragrance, and growth habit of once-blooming old roses with the repeat-blooming trait and color range of modern roses.

Along the central walk running west of the temple are planted climbers, All-American Rose Selections, and new introductions. South of that walk are a bed each of Hybrid Tea and Polyantha from 1867 to the 1910s and from the 1920s, a bed of Hybrid Tea, Floribunda, and Polyantha from the 1930s and two beds from the 1940s.

Along and near the west walk are beds of such specialty varieties as 'Peace', 'Ophelia', and 'Koster' roses and their sports on one side and plantings of wild roses and close hybrids from among more than two hundred known species on the other side.

'Altissimo' grows up the wall next to the tea room entrance.

BELOW: Mid-April in the rose garden: in the foreground Hybrid Tea roses are carefully graded according to height; the Hybrid Tea 'Snowbird' climbs up the arch in the background.

Violet trumpet vine (*Clytostoma callistegioides*) drapes itself behind a fine display of 'Givenchy', a richly scented Hybrid Tea rose named for the perfume maker.

• THE HERB GARDEN •
PLANTS FOR TASTE, SCENT, HEALTH, COLOR

In fragrances and flavors plants of the herb garden carry associations of foods, remedies, even flowers that go back to the roots of European culture. In the foreground is the brilliant red foliage of *Amaranthus* 'Molten Fire'. Its young leaves are used in salads and its seeds as grain. Hollyhocks (*Alcea rosea*) frame the tea room.

RIGHT: The autumn-flowering saffron crocus (*Crocus sativus*) is the source of the expensive spice saffron. The long red stigmas, twining through orange anthers, are cut, dried, and ground into a powder used for coloring and flavoring.

A HALF-ACRE FORMAL GARDEN planted to hundreds of kinds of herbs is tucked in a corner north of the rose garden and west of a patio restaurant remodeled from the estate's former billiard room and bowling alley. Installed in the 1950s, the garden reopened in its present form in 1976. Around an eighteenth-century German wrought-iron wellhead as a centerpiece, its separate beds group plants by use: culinary, salad herbs, teas, liqueur, medicinal, perfume and cosmetic, dye, strewing and sachet, and nosegays. Other beds hold such specialties as rosemaries, rugosa and sweetbrier roses, and lavender. Some herbs are perennials but many are annuals, flowering bulbs, or deciduous perennials that die back after their growing seasons, so much of the planting is changed seasonally.

Not a botanical but a colloquial term, "herb" has many definitions, none of them precise. Cooks often restrict the meaning to leaves of low-growing annual and perennial shrubs, but this garden takes a broader view: An herb is a plant whose parts—leaves, flowers, seeds, roots—have such special traditional uses as those noted above.

Like the Shakespeare garden, this one has living references to plant material recorded in the Library's collection of rare sixteenth- and seventeenth-century herbals. Those books contain botanical data, recipes, remedies, and other

The berries of poke weed (*Phytolacca americana*) become a rich black when mature. They were used as a source of both dye and ink by early American colonists. The plant is native to the eastern United States.

information about diet, health, and economic concerns in periods from medieval Europe to the settlement of North America. They describe many herbs obscure today but others familiar over the centuries.

The most widely known are those of American and European kitchen gardens, the likes of basil, laurel, marjoram, oregano, rosemary, sage, savory, tarragon, and thyme. Others are seasonings from other lands, increasingly enlarging the American food repertoire, such as Mexican oreganos and tarragons, Southeast Asian lemon grass, and Vietnamese cilantro. A salad-herb bed grows bronze fennel, nasturtium, parsley, and radicchio, while red-leaf basil in summer alternates with red-leaf lettuce in winter. A bed of confectionery herbs has horehound, a non-native widely naturalized in California; rose geranium, violets, and others. Two beds hold herbs that flavor wines and liqueurs and those whose leaves and other parts make teas and similar infusion drinks.

A bed of curative plants contains *Aloe vera,* an ancient specific for burns and other skin healing, as well as foxglove, larkspur, and *Catharanthus rosea,* from which chemicals used in treating leukemia can be extracted.

Plants were the main source for dyes until anilines and other chemicals came into use in the nineteenth century. Three seen here yield the primaries: red, blue, and yellow. A highly prized red came from madder. Leaves of woad (*Isatis tinctoria*) produced a blue that ancient Britons used not only for fabrics but to color their skin to intimidate even such sophisticated attackers as the invading Romans. Shades of yellow came from coreopsis, mullein, and several marigolds.

Four beds hold such cosmetic herbs as jojoba, used in shampoos, and such fragrance sources as gardenia, rose, and violet. One grows lavenders, roses, lemon verbena, and others used in potpourri and for petal-strewing on festive occasions. The charming idea of flowers as herbs appears in a tussie mussie, or nosegay, bed of blossom symbolisms. The floral code is an ancient tradition. In *Hamlet*, Shakespeare's audiences knew exactly what the fair Ophelia meant when she declaimed of pansies for thoughts, fennel and columbine for strength and folly, daisy and violet for innocence and modesty, and "There's rosemary, that's for remembrance."

The lavender-purple flowers of *Iris pallida* grace the herbaceous borders; the rhizomes are of value too in perfume and potpourri.

LEFT: Rectangular planting beds edged with bush germander (*Teucrium fruticans*) enclose plants from which medicinals can be extracted: foxglove (*Digitalis purpurea*) for the heart; feverfew (*Tanacetum parthenium*) for headaches; and *Aloe vera* for burns.

· THE JAPANESE GARDEN ·
ELEVEN CENTURIES OF TRADITION

The central feature of the Japanese garden is its canyon setting, meant to be strolled through, beside a small watercourse. It is designed to achieve a serene enviroment by shutting out the larger world. A moon bridge, so called because its arch and reflection approximate a full-moon circle, was built about 1912 by a Japanese craftsman. A graceful weeping willow overhangs the bridge, the swordlike foliage of the cycads guards its approach, and the golden dome of false cypress (*Chamaecyparis pisifera* 'Filifera Aurea') echoes its curve.

TWO STONE LION-DOGS guard a gate west of the rose garden beyond which the visitor descends into a nine-acre canyon garden. The understated entry gives little hint of the visual surprise to come: a descendant of a garden form traceable back to the tenth century in Japan, as interpreted in twentieth-century America.

With none of the broad spaces and geometry of European garden tradition, this is an informal space to stroll through. Like a good suspense novel, it tantalizes the stroller's curiosity to see what lies ahead while changing views and small landscape events slow his progress and divert his attention: an exquisitely pruned pine, a stone lantern, a miniature pagoda, a tiny shrine. The garden alternates open with intimate passages, structures and accessories with skillfully manipulated planting. Using sophisticated artifice to achieve simplicity and identification with nature, it projects a sense of tranquillity.

Begun in 1911, the Huntington's Japanese garden is one of America's oldest, most elaborate, and most gracefully matured. It has three parts. The main section is laid out as a private hill and stroll garden of a traditional house. A second section, added in 1968, is a walled compound containing a Zen-style garden and space for exhibiting bonsai (dwarfed trees). A third, added in 1970,

A tile-capped, earth-colored wall surrounds the Zen garden, an expanse of raked gravel meant to evoke the feeling of a flowing stream interrupted here and there by rocks and backed by trees and shrubs that suggest the far bank. The garden is virtually a reproduction of one in the main abbot's quarters in the celebrated Daitokuji temple in Kyoto.

OPPOSITE: Fall color comes to the Zen garden when fallen ginkgo leaves blanket the mondo grass planted as ground cover. The garden's plain stucco wall dramatizes both the forms and shadows of the trees.

is an informal three-acre garden of Asian plants, among them groves of golden and black bamboo and specimens of that hallmark tree of Japanese public and temple gardens, *Cryptomeria japonica.* In 1973 an Ikebana (flower arrangement) house was erected outside the public area, where members of the San Marino League, a garden support group, study the art of ikebana and create arrangements for display in the Japanese house.

Why an Oriental garden here amid so much Occidental tradition? Things Japanese were of widespread fascination in 1911, scarcely sixty years since Commodore Matthew Calbraith Perry opened the reclusive island empire to the outside world. In the next few decades, America marveled at Japanese gardens in expositions at Philadelphia, Chicago, San Francisco, and Saint Louis. Accounts by travelers to Japan, among them Edward S. Morse's 1885 *Japanese Homes and their Surroundings,* stirred the national imagination. In the early twentieth century that book influenced architects Greene and Greene of Pasadena, among others, to a whole new aesthetic of structure related to landscape. By then Japanese gardens had become the height of fashion for those who could afford them. Some were small tea-ceremony gardens, but in Los Angeles two of estate size were built in the Hollywood Hills in 1912, one of them by Myron Hunt, Huntington's own architect. Meanwhile, George T. Marsh, proprietor of the

Ornamental fruit trees lend seasonal color to Japanese gardens. In North Canyon the white-blossomed peach, *Prunus persica*, flowers early, a welcome harbinger of spring.

famous Japanese garden for the 1894 San Francisco exposition, had opened a commercial tea garden in Pasadena.

Thus there were ample information and even design guidelines available when Huntington and Hertrich decided on a Japanese garden. Among them was Josiah Conder's *Landscape Gardening in Japan,* published in 1893, a copy of which Hertrich owned. The chosen site was a former reservoir, overgrown with brush, on the west side of San Marino Ranch. Fortuitously, when Hertrich approached Marsh seeking to buy Oriental plants, he found the tea garden failing financially, so though the plants were not for sale the whole might be. Huntington, accustomed to buying entire libraries and collections, snapped up the ensemble. Then came a mighty effort at earth moving, dismantling Marsh's Japanese house, boxing plants and ornaments, planting, and reassembling the house—all in record time, as Hertrich described it. There was some urgency, for Huntington wanted the garden completed by the end of 1912, when his new residence was to be occupied. That likely anticipated his impending second marriage, in 1913, and he may have intended the garden as a present for his new bride, Arabella.

Hertrich designed the garden and supervised its creation. He hired a Japanese craftsman to build its centerpiece arch bridge and a tile-roofed pavilion for a bronze temple bell. He also imported more garden accessories from Japan and over the

Pine trees and a stone "firefly" lantern are mirrored in the surface of the pond.

years added other planting. By World War II the house and garden had fallen into disrepair, but in the 1950s they were refurbished, largely through the San Marino League's efforts, and opened in 1958.

Inside the garden gate, a wisteria-shaded path leads down to a terrace and a first stunning view over the main garden and across to the house. Spanned by its arch bridge and a smaller flat one, the pond is the focal point. Beside it are a large weeping willow (*Salix babylonica*), a dense cluster of cycads, and an evolving display of junipers and pines pruned and shaped for form. Pathways cross a lawn, climb the canyon sides, and follow a watercourse upstream.

Part of Japanese garden mystique is that the color green should be dominant, to induce a restful effect with none of the bright hues of, say, English flower gardens. That happens here, though a surprising amount of color plays its lesser, largely seasonal role. Much of it comes in spring from wisterias on three arbors. The pond has colorful water lilies and koi darting and flashing when sunlight catches their brilliant color. Mounding azaleas burst into bloom in late winter and spring on the slopes below the house. Flowering apricot, cherry, and peach trees blossom from January through April. Sweet olive (*Osmanthus fragrans*) trees near the house add a pervasive sweet fragrance with their tiny white spring flowers. Several varieties of Japanese maple (*Acer palmatum*) sport red leaves in spring and

Top: To the south of the bonsai court, a traditional meandering path is lined with coast live oaks, underplanted with the East Asian fern, *Microlepia strigosa*.

Bottom: The curving trunk of a mature oak and a young maple create the impression of an arch framing the pagoda. Winding paths in this stroll garden alternately reveal and conceal views.

Left: The winter sun highlights the flossy green heads of *Cycas revoluta* and blossoms of the flowering peach (*Prunus persica* 'Icicle') by the arch bridge.

Japanese maples (*Acer palmatum*), among the most decorative of foliage trees, bring a touch of amber to the garden in the fall and a rich crimson in spring when new leaves appear.

summer and yellow in the fall—when the Zen garden's maidenhair trees (*Ginkgo biloba*) turn a bright yellow.

Pines, especially those native to Japan, are important features of Japanese gardens. Handsomely groomed Japanese red pines (*Pinus densiflora*), including the 'Dragon's Eye' cultivar, are prominent around the stroll garden in single and multiple-trunked, weeping, and other forms. Many Japanese black pines (*P. thunbergii*), pruned and shaped as meticulously as bonsai, stand out on an island in the pond by a stone Buddha, on central garden slopes, and on slopes around the Zen garden.

The canyon's west side is topped by a five-room nineteenth-century house from Japan. Following the Chinese precedent, the Japanese learned to orient the interiors of their houses to gardens outside. This one shows how traditional features accomplish that goal: the engawa, a veranda or floor-level platform that affords room-to-room passage outside; shoji, sliding exterior panels covered in translucent rice paper that close off rooms or open them to the garden; and amado, outer sliding shutters that close the house at night or during rain. Its interior features include fusuma, sliding wall panels that open up or close off rooms; tokonoma, alcoves for the reverential display of a flower arrangement, scroll, and other objects; and tatami, rush-covered straw floor mats in modular sizes that determine the size and shape of rooms. Though this is not a tea house, a brazier and a tray with a teapot and cups in one room imply the tea ceremony. One of the most important rituals in Japanese culture, it is traditionally a party of a host and four guests.

South of the house, a wisteria arbor shades a display of suiseki, water-worn rocks in the shape of miniature landscapes. Beyond, steps lead down to a zigzag bridge across a dry landscape of stones that suggest moving water in a creek. The idea that the purpose of a zigzag bridge is to thwart pursuit by evil spirits, known to

This traditional house was imported from Japan and reconstructed early this century, first in Pasadena and a few years later at the Huntington.

This zigzag bridge crosses a dry streambed, where pebbles evoke the presence of rushing water. On the slope are Japanese black pine *(Pinus thunbergii)*; to the left of the steps tower two Canary Islands pines: and clipped shrubs on either side of the steps are the Japanese *Pittosporum tobira.*

LEFT: This is an island-shaped stone, an example of the Japanese art of *suiseki*, natural stones that suggest in miniature such landforms as a distant mountain or an island in the sea.

RIGHT: This European olive, a representative of the ancient art of bonsai, is carefully pruned to keep it miniature and shapely.

travel only in straight lines, may be apocryphal but the form is also a practical way to span a stream with only one set of supports in the center.

Then steps ascend a slope planted to black pines, junipers, and tobira (*Pittosporum tobira*) pruned in rounded shapes, past two huge Canary Island pines (*Pinus canariensis*) to a rectangular garden inside a tile-capped wall. Here on one side of a central slate walk an expansive dry garden of raked gravel, a few stones, and background shrubbery makes a stylized, abstract reference to islands in a moving stream. Rather than an illusion, this is considered to be a distillation of the atmosphere that rock and water might produce in that elusive, contemplative spirit known as Zen. On the other side, ginkgoes over a ground cover of mondo grass (*Ophiopogon japonicus*) similarly evoke a forest. Past a far gate, a small court displays bonsai, those often old dwarfed trees in shallow containers kept shaped by continual pruning of leaves, branches, and roots. At one end of the court a path leads down through a shady forest of giant bamboo and tall trees, mostly from Asia, to the garden's lower exit.

OPPOSITE: A looped fringe of wisteria (*Wisteria sinesis* and *W. floribunda*) softens the geometry of the Japanese house. The raised floor is covered with tatami, mats made of compressed rice straw with a woven rush surface.

PAGES 128-129: Next to the Japanese garden, the subtropical garden displays such beauties as *Erythrina falcata*.

In The Gardens, Part III

THE NEWER GARDENS

THE AUSTRALIAN GARDEN
TREES THAT CHANGED THE FACE OF CALIFORNIA

Eucalyptus dominate in height and in number the Australian garden. The oldest specimens, like the tall and elegant *Eucalyptus citriodora* at left, were planted in the 1940s. The Huntington has nearly one-fourth of the 700 eucalyptus species, showing their remarkable range of height and growth habit, leaf form and color, bark texture and color, often long-lasting flowers, and sculptural fruits and seed pods.

A member of the daisy family *(Asteraceae)*, Billy Buttons *(Craspedia globosa)* is a good cut or dried flower.

THE LINEAR AUSTRALIAN garden covers five acres of level terrace ground directly below and parallel to the subtropical garden. Stretching from the lily pond area to the Japanese garden, this partly wild, partly parklike passage of trees and shrubs mostly native to Australia is dominated by eucalyptus. A tramp through its meadows and groves makes a pleasant contrast with a walk on paths through the more manicured gardens on the hilltop above.

Like palms, eucalyptus early on became a California tradition. Palms may be the Southern California symbol, but no trees more drastically altered the face of the state than eucalyptus, the most widely planted of all imports. Eucalyptus enthusiasm began in the nineteenth century and peaked in a craze of planting for timber and railroad ties even as Henry Huntington was developing his gardens. Less interested in mass than in selective planting, he put in only a few, among them lemon-scented gums *(Eucalyptus citriodora)* near today's patio restaurant and next to his mausoleum.

Widespread interest continued, and in 1928, after Huntington's death, William Hertrich had some rare and interesting eucalyptus installed in a northeast part of the property now devoted to visitor parking. In 1943 a thousand trees

RIGHT: Magenta pink flower clusters form at branch tips on the pink rice flower (*Pimelea ferruginea*) from Australia and New Zealand.

OPPOSITE: *Eucalyptus deglupta*, from the Philippines, is prized for its multicolored bark and leaves that are aromatic when crushed.

supplied by the U.S. Department of Agriculture were planted in close rows on the site of a former orange grove south of the estate house. Later all but fifty were taken out. Those veterans form the nucleus of today's Australian garden, opened to the public in 1964 after many other plants had been added.

Eucalyptus is the most numerous genus in Australia, which has some 700 species. The garden has 160 species, both trees and shrubs. Placed among them are acacias, callistemons, cassias, grevilleas, melaleucas, and others from the island continent's semiarid regions. At the garden's eastern end a collection of New Zealand natives now taking shape specializes in phormiums and other plants notable as landscape subjects more for leaf and structure than for flowers.

Many trees have grown to monumental height, among them the tall, straight,

From winter through summer, but especially in spring, the weeping branches of *Eucalyptus woodwardii*—an endangered species from western Australia—seem to droop under the weight of the lemon-yellow flowers as they emerge from button-tight buds. It has one of the most beautiful flowers of all the smaller, shrubby eucalyptus called mallees.

fast-growing Sydney blue gum (*E. saligna*). A young specimen mountain ash (*E. regnans*), the tallest eucalyptus and tallest of all hardwood (flowering) trees, grows in the southeast part of the garden. Set out among the trees, many other smaller, shrubby eucalyptus called mallees are famous for showy flowers of yellow, pink, and red.

Here too are acacias, of which Australia also has hundreds of species. In the garden they stand out for their winter and spring bloom, especially *Acacia podalyriifolia*, with masses of tiny yellow blossoms and silvery foliage, and golden wattle (*A. pycnantha*), considered the national flower, covered with clusters of sweet-scented yellow flowers. Specimens of silky oak (*Grevillea robusta*), with orange-yellow flowers in late spring, rival some of the old eucalyptus trees in size, height, and girth. Brachychitons are seen here as large specimens alone and in a grove of fifteen or so, and as younger trees with distinctive bottle-shaped trunks. Melaleucas are represented by the snow-in-summer (*Melaleuca linariifolia*), snowy white in spring with flowers on spreading branches above a stocky trunk; the chenille honey-myrtle (*M. huegelii*), with spikes of white flowers in late spring, from western Australia; and small species. A young gum myrtle (*Angophora costata*) will grow to a tall tree that resembles eucalyptus in size, habit, and even leaves.

The large trees are seen from the road along the south side or a path along the north, while a walk through the wilder middle reveals more variety. Plants of one of Australia's four cycad genera, *Macrozamia,* are seen in the garden. The spear lily (*Doryanthes palmeri*) and gymea lily (*D. excelsa*) are spectacular in spring with six-foot or longer flower stalks and large red flowers above tall sword-leaf clumps. Two plants have copious lavender flowers, the fragrant-leaved mint bush (*Prostanthera rotundifolia*), and blue hibiscus (*Alyogyne huegelii*). Standing out among fifteen species of cassias are silver cassia (*C. artemisioides*) and desert cassia (*C. nemophila*); both have bright yellow flowers and silvery, feathery leaves. The oddest plant of all is red and green kangaroo paws (*Anigozanthos manglesii*), with truly bizarre, fuzzy green flowers atop thin red stalks; other species have red or yellow flowers.

Mottlecah (*Eucalyptus macrocarpa*) has the largest flowers and fruits of all eucalyptus. As the red stamens appear, the hard white bud cap (top), actually the fused petals, falls off. Fluffy red stamens push apart the gray-green leaves ranged along sprawling horizontal branches (bottom).

ABOVE: Acacias are the second most numerous species in the Australian garden. Pearl acacia (*Acacia podalyriifolia*) blooms in January. The seed pods, which follow the welcome yellow mimosa flowers, are the same blue-green as the feathery leaf sprays.

RIGHT: Gardeners appreciate the fernlike leaves and amber-colored spring and summer flowers of the fast-growing silky oak (*Grevillea robusta*), and furniture makers prize its wood.

OPPOSITE: *Acacia drummondii* ssp. *elegans*, a small ornamental shrub covered profusely in spring with cylindrical yellow flowers, is also valued by gardeners.

The stiff sword-shaped leaves of New Zealand flax are useful to define a border, mark the corner of a pond, or announce a flight of steps. Here, planted informally in a woodland, emphasized by streaming sunlight, is the lower-growing *Phormium* 'Apricot Queen' and *P.* 'Sundowner'.

Anigozanthos flavidus is known as kangaroo paws because of the clawlike shape of its tubular flowers. Carried on stems up to six feet tall, these are usually greenish yellow, sometimes yellow, orange, or red. Although many species bloom in winter this one is summer-flowering.

THE SUBTROPICAL GARDEN
PLANTS FROM MEDITERRANEAN CLIMATES

Cassia spectabilis carries its large, long-lasting flowers in August and September. In the foreground, the yellow Acapulco daisy (*Zinnia maritima*), a Huntington introduction, blooms all year unless caught by a frost.

RIGHT: The long seedpods of the golden trumpet tree (*Tabebuia chrysotricha*) split open when mature to release winged seeds.

I N THIS four-acre garden, "subtropical" refers to plants that can survive a light frost but are likely to perish when the temperature drops well below freezing. Occupying the estate's warmest spot and best protected from cold, the garden tests plants whose long-term performance in the San Gabriel Valley climate remains to be learned. On the slope below the Huntington Gallery and rose garden and above the Australian garden, it extends east from the Japanese garden to the jungle garden.

The garden displays bulbs, shrubs, trees, and non-succulent plants from sub-tropical regions. With such diversity it lacks the orderly appearance of the more structured gardens and its unfamiliar species make it of special interest to botanists and other plant enthusiasts. Represented are the Mediterranean region and areas of similar climate in South Africa, South America, and islands in the eastern North Atlantic—particularly the Canaries—as well as warm-temperate and subtropical Africa, Southeast Asia, and the Americas. Not represented are Mediterranean-climate parts of Australia, which have their own adjacent garden, or California, seen in other Southern California botanical gardens devoted to native plants.

Many plants here came from Huntington staff expeditions to Mexico, others were sent by plant explorers from around the world. While the planting arrangement

LEFT: The beautiful, long angels' trumpets of *Brugmansia versicolor* are white flushed with peach.

RIGHT: The showy flowers of canna, which is related to ginger, bloom from spring to fall.

OPPOSITE: The path winding to the garden pavilion designed by the celebrated landscape architect Thomas Church is lined with shrubs and bulbs. In the foreground are the blue flowers of the South African native, agapanthus, which are synonymous with summer in Southern California.

has been informal, some specimens are gradually being relocated by geographic origin: Mexico toward the eastern end, southern Africa in the center, Mediterranean in the west.

Along the several paths lacing the slope the most prominent—and permanent—features are flowering trees. Cassias put forth yellow flowers in late summer and fall. A large pink silk floss tree (*Chorisia speciosa*), unusual for having no prickles on its green bark, blooms in late fall. Winter brings out bloom on coral trees of several species, one of them a spectacular hybrid, *Erythrina ×sykesii*. Skyrocket flower (*Leucospermum reflexum*), one of few South African proteas that are reliable in Southern California, blooms for three months from February on.

Spring sees flowers on jacaranda in blue, orchid tree (*Bauhinia blakeana*) in purplish pink, and a landmark cape chestnut (*Calodendrum capense*) in pink. Golden trumpet tree (*Tabebuia chrysotricha*) blooms on bare branches in early spring, then leaves appear. *Wigandia urens*, with deep purple spring flowers and huge leaves, stands out like a beacon. The more shrublike tree dahlia (*Dahlia imperialis* 'Plena') has pinkish purple flowers.

This garden has more color on the ground than most. It comes from dozens of South African bulbs in bloom at different times of the year and such ground covers as *Gazania uniflora*; cape weed (*Arctotheca calendula*), a vigorous yellow daisy; and *Trachelospermum asiaticum*, related to star jasmine.

Among several Huntington introductions are the Acapulco daisy (*Zinnia maritima*), a ground cover; the 'Irving Cantor' agapanthus, with low foliage and blue flowers on three-foot stalks; and *Salvia leucantha* 'Purple Velvet', part of a remarkable salvia collection.

A little off the paths grow such rarities as mountain lobelias, found only in the highest mountains of East Africa, and such curiosities as the canary bird bush (*Crotalaria agatiflora*), the daisy tree (*Podachaenium eminens*), and *Boehmeria nivea*, from which ramie fiber is extracted.

OPPOSITE: A spectacular burst of color occurs in October and November with the flowering of *Gladiolus dalenii*, from which our modern hybrids derive.

ABOVE: Pink powder puff (*Calliandra haemato-cephala*) is a winter-blooming evergreen or semideciduous shrub.

BOTTOM RIGHT: The marma-lade bush (*Streptosolen jamesonii*), an evergreen shrub, can reach eleven feet, displays its tubular orange-red flowers in the subtropical garden.

BOTTOM LEFT: Chameleonlike flowers of the shrubby vine *Combretum fruticosum* turn from chartreuse to yellow to orange-red.

THE JUNGLE GARDEN
TROPICAL FANTASY FOR A SUBTROPICAL CLIMATE

A fretwork of leaves dapples the cooling waterfall at the heart of the four-acre jungle garden. The *Ficus auriculata* on the right was already here, part of a forest of ornamental fig trees, when this garden was opened in 1979. On the left Australian tree fern (*Cyathea cooperi*) with its lacy foliage relishes the damp atmosphere, and Abyssinian banana (*Ensete ventricosum*) raises high its tattered leaves in the background.

RIGHT: Hibiscus flowers are solitary beauties, their paper-thin petals a marvel of subtle shading and a reminder of the tropics.

RATHER THAN TRY to reproduce an equatorial rain forest, this four-acre garden creates tropical ambience with species that thrive in a subtropical climate. Suggesting the riotous growth of the hot and humid low latitudes, this multilayered environment has a high forest canopy, understory trees and shrubs, vines climbing tree trunks, leaves of often gargantuan proportions, but especially plants that people associate with the tropics. Grand old evergreen fig trees (*Ficus thonningii*) and Chinese wingnut (*Pterocarya stenoptera*) spread shade over bamboos, bananas, bromeliads, cycads, ferns, gingers, palms, and philodendrons. A waterfall and a stream tumbling down through this cool, dim retreat encourage moisture-loving plants. Located on a south-facing hillside with the same warm microclimate that benefits the adjacent subtropical garden, this informal, wild landscape also merges with the palm garden, visually borrowing its tropical effect.

The stream, completed in 1978, fills the lily ponds below. Near the ponds rears the most remarkable tree in the jungle garden if not on the entire estate, an ombú (*Phytolacca dioica*). Fast-growing, tall (to sixty feet) and nearly as wide, native to the plains of Argentina, it is that nation's unofficial emblem. A gnarled old specimen grown from seed received from the Buenos Aires Botanical

147

Assertive leaves of the elephant ear plant (*Xanthosoma*) and split-leaf philodendron (*Monstera deliciosa*) clothe the intricate root system of an ornamental fig tree. The edible fruits of *Monstera* taste of pineapple and banana.

OPPOSITE: A walk, wild in feeling, from the waterfall to the Huntington Gallery is lined with *Pritchardia* palms and old specimens of the cycad *Encephalartos ferox*. In spring and late summer (first half of September) the small red flowers of *Billbergia* 'Huntington' appear. Later in the year the nectar-producing flowers of bird of paradise (*Strelitzia reginae*) and the luminous red flowers of the coral tree brighten the walk.

Garden in 1912, this ombú has an enormous swollen base, part of an above-ground root system that evolved for water storage as an adaptation to grass fires, winds, and scarce rainfall of the pampas.

The jungle garden's bananas and their relatives the heliconias and strelitzias rank high for tropical image. Among banana species grown as ornamentals is *Musa ornata*, with showy pink bracts. A relative of the banana, traveler's tree (*Ravenala madagascariensis*) may derive its name from a tendency to orient itself under natural conditions on a north-south axis, supposedly acting as a compass to travelers.

In the canopy shade grow several species of ficus, including bo tree (*Ficus religiosa*), native to India and Southeast Asia, sacred as the tree under which the Buddha sat when he received enlightenment. Podocarpus, unusual broadleaf conifers mostly from the Southern Hemisphere, are represented here by one from Mexico, *Podocarpus reichei*. The coffee tree (*Coffea arabica*), native to Africa, grows here too, though at risk from frost.

The garden has two of the few New World bamboos in cultivation, Mexican weeping bamboo (*Otatea acuminata*) up near the waterfall, and the rare *Chusquea coronalis*, near the eastern edge. Another bamboo, the rare *Dendrocalamus asper* also grows near the waterfall.

To take advantage of the warm microclimate, new palms are planted on both sides of the boundary with the palm garden next door. Seen along the paved walk

Strong forms of the conifer collection loom beyond the edge of the jungle garden. Lining the sloping path are clumps of mondo grass (*Ophiopogon*), pierced by the striking yellow-green and pink-green foliage of anthuriums and bromeliads.

OPPOSITE: The lower side of the garden, close to the lily ponds, enjoys a warm, sheltered microclimate. Here is the strawberry snowball tree (*Dombeya cacuminum*), a native of Madagascar, whose lovely scented flowers are sometimes hidden by luxuriant leaves.

are *Chamaedorea elatior*, a rare climbing palm, *Pritchardia*, native to Hawaii and other Pacific islands, and fishtail palm (*Caryota urens*). Cycads also contribute lacy, palmlike foliage, as do Australian tree ferns (*Cyathea cooperi*), with distinctive silver-dollar leaf scars on the trunk.

Ginger family genera *Hedychium* and *Alpinia* contribute bright colors in the dark forest, and split-leaf philodendron (*Monstera deliciosa*) and other plants add bold leaf forms. So do *Philodendron* 'Evansii', an arborescent or treelike plant, and a great-leaved elephant ear (*Xanthosoma*), near the waterfall. Philodendrons also grow in trees, though not as parasites, and some survive there as epiphytes after their connecting stem dies away. Other epiphytes that cling to trees include ferns, orchids, and the cacti *Rhipsalis* and *Epiphyllum*. Among lianas, large woody vines that hang from trees, are the chestnut vine (*Tetrastigma voinieranum*) from Laos and *Aristolochia braziliensis* from Brazil. Forest-floor plants include such terrestrial bromeliads as pineapple (*Ananas bracteatus*) and species of *Pitcairnia*. Bromeliads that are naturally epiphytes are also used as ground covers here: *Aechmea, Billbergia, Nidularium, and Neoregelia.*

Many bromeliads are stemless perennials with colorful bracts and showy flowers that often last for months. A sampling of the species in the jungle garden is pictured here.

OPPOSITE PAGE

LEFT: The brightly colored bracts of *Portea petropolitana* compose its long-lasting four-foot flower stalks, which project from stiff swordlike leaves.

TOP RIGHT: Easy to grow, *Billbergia pyramidalis* takes its name from its pyramid-shaped flower spike.

BOTTOM RIGHT: The flowers of *Hohenbergia stellata* last for months. Its name refers to the starlike form of the compact spikes of the flowers.

THIS PAGE

TOP LEFT: The rose-colored torch of *Quesnelia arvensis* is studded with tiny dark blue and white flowers.

TOP RIGHT: *Aechmea fasciata* is the most popular and widely planted bromeliad in the world. Its rosy pink flower bracts have pale blue flowers that change to deep rose.

BOTTOM: The inflorescence of this *Aechmea ornata* is attractive even though it is almost through bloom-ing—a few light pink flowers are still visible on top.

THE ENTRANCE
AND PARKING GARDENS
A VISITORS PAVILION AND PARKING AMID PLANTING

Tabebuia chrysotricha extends its yellow trumpet flowers in the March sunshine above an underplanting of lavender.

ARRIVING VISITORS first park in a lot softened by islands of greenery in the blacktop and plantings of background trees and shrubs. At twenty-five acres, a sixth of the Huntington's public area, this garden holds twelve hundred cars with little of the wasteland effect of most parking lots. South of the lot a small, more formal green garden is planted with podocarpus, coast redwood, sycamore, and a line of eight carrotwood trees (*Cupaniopsis anacardioides*). It serves as prelude, inviting visitors into an entrance pavilion. That in turn acts as a kind of vestibule to a broad terrace and a small plaza beyond, planted to bright seasonal flowers, flowering trees, and lofty palms, overlooking the grand sweep of gardens to the south.

In the northeast part of the estate, the parking lot was at first a kitchen garden, a place for growing fruits and vegetables and raising poultry, and also a test plot for

This escutcheon atop a pair of wrought iron gates at the seldom-used Euston Road entrance bears the arms of the Carew family. The gates came from Beddington Park, Surrey, England, where they were erected by Sir Nicholas Carew in 1714.

Chinese date, grapefruit, pistachio nut, persimmon, and other exotic fruits. In 1908 Henry Huntington began planting in the northern half of this area what is believed to be California's first commercial avocado grove. Some gnarled veterans with massive trunks remain, still producing fruit. The arboretum aspect of avocado cultivation continues as new trees are planted in a growing collection of historical cultivars north of the parking lot

As the year advances, the colors become richer. The maidenhair tree (*Ginkgo biloba*) assumes its yellow autumn foliage, offset by the thick cream plumes of pampas grass (*Cortaderia selloana*). In the foreground the Chinese plumbago (*Ceratostigma willmottianum*) has slate blue flowers and red-tinted autumn leaves.

RIGHT: A crape myrtle blooms behind the nineteenth-century Borghese vase, a copy of a Greek bronze urn. Every year half a million visitors come down these steps outside the entrance pavilion to explore the gardens.

In summer the showy flowers of the crape myrtle (*Lagerstroemia indica*) catch the eye; its decorative peeling bark is an additional attraction.

in a cooperative effort with the California Avocado Society, which deposited its records at the Huntington in 1992.

Later the ground south of the avocado grove became a test planting of rare eucalyptus and other Australian trees, which remained in place long enough to achieve mature size. In 1966 the area was opened to the public as a general arboretum, and in 1979 it became a parking lot related to the pavilion, which opened two years later.

Thirty-two islands and peripheral dividers that define the lot's parking bays and lanes are planted to mostly drought-tolerant trees, shrubs, and ground covers. They preserve specimens from the former avocado orchard, fruit tree planting, and arboretum, and also display new material from a continuing planting program. The central section of the lot contains the Australians, the large eucalyptus and such others recently planted as bottlebrush (*Callistemon*), grevillea, and senna (*Cassia*). Another veteran from the old test planting is cork oak (*Quercus suber*) from the Mediterranean.

A central walkway in the southern third turns spectacular in spring with the blue-violet bloom of jacaranda trees (*Jacaranda mimosifolia*) that line it on both sides. The garden has two Huntington tree introductions, strawberry snowball (*Dombeya cacuminum*) and flowering cherry (*Prunus serrulata* 'Pink Cloud'). Other flowering trees and their colors are silk tree (*Albizia julibrissin*), pink; orchid tree (*Bauhinia blakeana*), varicolored pink; fringe tree (*Chionanthus retusa*) and snowdrop tree (*Halesia diptera*), both white; princess tree (*Paulownia tomentosa*), blue; tulip tree (*Liriodendron tulipifera*), pale green; and the recently planted golden trumpet tree (*Tabebuia chrysotricha*), yellow.

Among other more recent tree plantings are tipu tree (*Tipuana tipu*) from South America, maidenhair tree (*Ginkgo biloba*), and several cultivars of sweet gum (*Liquidambar styraciflua*).

• LANDMARK TREES •
LARGE, RARE, COLORFUL, OLD

Trees are a vital part of this landscape, interesting as individuals and awe-inspiring as a collection. Gracing the gardens are more than a thousand native oak trees and examples of more than forty introduced oak species, many notable for their rarity or unusual size. The magnificent specimen at left is a deciduous English oak, *Quercus robur*, uncommon in California.

The Montezuma cypress (*Taxodium mucronatum*) at right was grown from seed from Mexico.

OVER THE DECADES, many of the Huntington's trees have become landmarks for unusual size, age, rarity locally or nationally, or for having notable flowering or other features. Some had a head start toward landmark status, for Henry Huntington encouraged William Hertrich to bring in specimens as mature as he could find—or move. Hertrich devotes a section of his *Personal Recollections* to adventures in moving and planting large trees without the use of modern equipment. Indeed, his pioneering experience with such transplanting was put to practical use in large-scale landscaping for the 1915 expositions in San Diego and San Francisco. Today the Huntington has an assemblage of landmark trees, some of them the only examples or the largest of their kind, some the parents of trees in other gardens.

South of the entry pavilion, several specimens of silk floss tree (*Chorisia speciosa*), Queensland kauri (*Agathis robusta*), and Senegal date palm (*Phoenix reclinata*) serve as introduction to garden delights to come. In the lawn beyond, three of the most impressive oak specimens on the grounds are English oak (*Quercus robur*).

Some of the Huntington Gallery's landmark trees include a large specimen *Talauma hodgsonii*,

The beautiful lemon-scented gum (*Eucalyptus citriodora*) was a favorite of Henry Huntington's, who asked that this species be planted by his mausoleum. The smooth white bark emphasizes the tree's elegant structure from afar, and close up, it has a pleasing, powdery-white surface—until it sheds its bark every summer. Crush the leaves and the lemon fragrance is unmistakable.

RIGHT: The evergreen cork oak (*Quercus suber*) will grow as high as it is wide. Its bark is exceptionally thick and able to regenerate after cork has been cut from a new layer that forms each year. In commercial groves, the bark is harvested for cork after twelve years' growth.

a magnolia relative from the Himalayas, and a tall, white-flowered *Chorisia insignis,* from the upper Amazon basin in Peru, adjoining the loggia, and a massive, century-old English yew (*Taxus baccata*) south of it. Other large flowering trees are seen south and east of the building: flame tree (*Brachychiton acerifolius*) and Moreton Bay chestnut (*Castanospermum australe*) from Australia, Kaffirboom coral trees (*Erythrina caffra*) from South Africa, and pink snowball tree (*Dombeya × cayeuxii*).

The rose garden has three monumental Montezuma cypress (*Taxodium mucronatum),* trees grown from seed Hertrich collected from Chapultepec Park in Mexico City in 1912. A giant of this species in the state of Oaxaca is Mexico's largest and probably oldest tree. Two other conifers stand out for great size in the rose garden. One is a Queensland kauri (*Agathis robusta*), planted in the 1890s by the San Marino Ranch's previous owner then transplanted in 1908 despite its forty–foot height to make way for the mansion. The other is an umbrella-shaped Italian stone pine (*Pinus pinea*) located near the herb garden but a prominent part of the rose garden skyline. Also near the restaurant is *Magnolia delavayi*, one of the few evergreen magnolias from Asia hardy in Southern California.

In the subtropical garden grows an Italian stone pine of equal stature but better sited to reveal the full drama of its form. Among the subtropical garden's large collection of flowering trees an old and distinguished cape chestnut (*Calodendrum capense*) stands out when covered with blossoms in the spring.

Several brachychitons are in the Australian garden: *B. rupestris*, identified by its

The conifer collection is on the west side of the lily ponds. The importance of trees in the landscape can be seen by comparing this recent picture with the historic photograph on page 32.

swollen trunk, and *B. discolor*, planted in a grove, striking for great quantities of rosy-pink bell-shaped blossoms. The Australian garden also has many significant eucalyptus, some unusual for great size, especially *Eucalyptus tereticornis*.

The conifer collection west of the lily ponds has several landmark trees. Two are known to be the largest specimens in the United States, *Keteleeria davidiana*, a pine relative from China planted in 1926, and Taiwania (*Taiwania cryptomerioides*), a cypress relative native to Taiwan and China planted in 1908. Perhaps the handsomest in the grove is a Kashmir cypress (*Cupressus cashmeriana*), with pendulous needles and elegant form. Growing here are the largest examples on the grounds of deodar cedar (*Cedrus deodara*) from the Himalayas, and southern magnolia (*Magnolia grandiflora*)—though the tallest specimen grows at the southwest corner of the Huntington Gallery. The rare Guadalupe cypress (*Cupressus guadalupensis*), endemic to Guadalupe Island off the coast of Mexico, also grows here. Also of note are araucarias, tall Southern Hemisphere cone-bearing trees native to South America, Pacific islands, and Australia; two tall species here are *A. cunninghamii* from Australia and *A. columnaris* from New Caledonia.

A landmark at the edge of the jungle garden is one of the oddest trees in any of the gardens, an ombú (*Phytolacca dioica*) from Argentina, a gnarled old veteran with a giant swollen base. Also in the jungle garden, three *Ficus macrophylla* var. *columnaris*, are notable for their height.

In the palm garden, the rare Chilean wine palm (*Jubaea chilensis*) is still one of

LEFT: Deodar cedar (*Cedrus deodara*), a fast-growing forest tree in the Himalayas, can reach one hundred feet in height and spreads its branches in a skirt forty feet wide at ground level. This specimen, grown from seed by Hertrich, extends its hanging branches along Deodar Lane.

RIGHT: The distinctive cones of *C. deodara*, borne high up on the crown, are a sign of maturity: trees less than forty years old seldom bear cones. The blue-gray leaves change to a dark green as they age.

the most impressive, but a landmark for rarity is a windmill palm (*Trachycarpus martianus*), a Himalayan native, the most mature specimen in a public garden in Southern California.

Not a tree, perhaps, but the desert garden's patriarch cactus, a *Cereus xanthocarpus*, mature when originally planted, weighs in at an estimated fifteen tons. Taller and more treelike, an outstanding specimen dragon tree (*Dracaena draco*) has grown to such heroic size its heavy limbs need steel-pipe props.

In the northern area, a blue Atlas cedar (*Cedrus atlantica* 'Glauca') at the junction of Deodar Lane and Mausoleum Road is one of the largest in Southern California and the parent, through cuttings, of most blue Atlas cedars distributed by nurseries in the region. And Mausoleum Road is lined with a particularly fine planting of Guadalupe Island palm (*Brahea edulis*).

Among parking lot trees, one landmark is a *Calocedrus formosana*, a rare conifer from Taiwan, in the southwest section. Many of the larger specimen trees are eucalyptus species from a 1920s test plot and avocados from an earlier planting.

The ombú tree (*Phytolacca dioica*), planted in 1914, has a thick crown of leathery leaves. Its enlarged base is a water-storage device that evolved for survival on the plains of its native Argentina. It can reach sixty feet in height and nearly that in width.

TOP LEFT: White silk floss tree (*Chorisia insignis*), native to Peru, guards the entrance to the Huntington Gallery. The prickle-studded, swollen, bottle-shaped trunk has evolved for water storage in dry periods. Large white flowers appear on bare branches before the leaves emerge.

TOP RIGHT: The seeds of *C. insignis* are protected by a fiber similar to kapok, (*Ceiba pentandra*) from a tree in the same family as chorisia.

BOTTOM RIGHT: Chorisias usually have smooth trunks, thickly studded with prickles that may disappear with age.

The silk floss tree (*C. speciosa*) is native to Brazil and Argentina. Its late-summer flowers range in color from pale to deep pink.

ABOVE: Although the rain forest is its natural habitat, the flame tree (*Brachychiton acerifolius*) has adapted well to the dry California climate. It has glossy foliage and brilliant orange-red flower trusses.

LEFT: Tall columnar landmark, this splendid Queensland kauri (*Agathis robusta*) was planted in the 1890s, then transferred as a mature forty-foot tree to make way for the new Huntington mansion.

OPPOSITE: Montezuma cypress (*Taxodium mucronatum*) is evergreen in mild climates, although the old foliage falls in early spring as the new leaves appear. Its weeping branches and slender leaves make a graceful tree that may attain great stature with age.

TOP: *Cassia spectabilis* carries a generous crop of bright yellow flowers at the end of summer, in August and September.

BELOW AND RIGHT: The emergence of the first delicate and scented blooms of the deciduous magnolia is a welcome winter sight. The Huntington collection is comprised mostly of small trees, with flowers ranging in color from white through pink to purple-pink.

The Library's smooth Ionic columns are the backdrop for this coral tree (*Erythrina coralloides*). From spring until winter its red flowers are visited regularly by humming-birds.

A young specimen of the Alexandra palm (*Archon-tophoenix alexandrae*) has a dramatically striped trunk. Growing normally in rain forest conditions, it will reach ninety-five feet. A native of Australia, it was named for Princess Alexandra of Denmark.

OPPOSITE: *Ficus macrophylla* var.*columnaris* is a native of Lord Howe Island, off the coast of Australia. This is a good example of buttress trunks found in many tropical trees.

• ADORNMENTS TO THE GARDENS •
TEMPLES, SCULPTURE, FOUNTAINS, ACCESSORIES

Floribunda roses 'French Lace' surround the Temple of Love in the rose garden. A structure from a dependency of the Palace of Versailles, it was possibly designed by Jacques Ange Gabriel (1710–1782) for the Versailles gardens. Inside is the eighteenth-century stone sculpture *Love the Captive of Youth,* by or after Louis-Simon Boizot (1743–1809).

IN THE GRAND European tradition, fountains, sculpture, and accessories embellish the Huntington gardens. More than three hundred such objects in stone, metal, and wood, many of them eighteenth century, some more recent, serve as focal points, accents, or merely adornments. Four are in the tempietto, or "small temple" genre, of which the largest and most imposing is the Huntington mausoleum, reminiscent in form and scale of Donato Bramante's celebrated 1502 Tempietto in Rome—so influential a Renaissance structure that it undoubtedly was part of architect John Pope's education. Three smaller, light-hearted temples contain sculptures on the theme of love: at the south end of the North Vista, in the rose garden, and on the south lawn below the Huntington Gallery. A bronze Saint Francis stands beside the lily ponds. The finest sculptures are placed near the Library and the Huntington Gallery, while the most impressive display is an array of stone figures lining the North Vista. Large ornaments include two elaborate fountains, in front of the Library and at the focal point of the North Vista. Smaller fountains are seen at the Irvine Orientation Center's entrance and on the Huntington Gallery's north and west sides.

Among seventy other garden accessories, in the Japanese garden are such special touches as an arch bridge and a classical zigzag bridge, a bell pavilion, and stone lanterns, statues, pagodas, and votary tablets. More than a hundred benches are disposed throughout the gardens in carefully chosen, agreeable spots, some of carved marble but mostly of wood, many of them memorials to individuals. The inventory of garden adornments also includes carved or molded stone and ceramic urns and planters, freestanding pillars, ornamental gates, and pavilions for shade and rest.

TOP LEFT: The stunning red-flowering spear lily (*Doryanthes palmeri*) blooms at the foot of Persephone, daughter of Zeus and Demeter, installed near the north door of the Huntington Gallery. The 1750 stone statue is from the Hofburg (the imperial palace) in Vienna.

TOP RIGHT: Depicted with flowing hair and a thick mustache, this limestone Mars, part of the North Vista statuary, is framed by *Camellia japonica* 'Glen 40 Variegated'.

BOTTOM LEFT: Half dog, half lion, Koma-inu is one of a pair of stone guardian figures. Pairs are stationed at the main and south entrances to the Japanese garden, here as in Japan.

BOTTOM RIGHT: With his game slung over his shoulder, a young boy and his dog return from the hunt. The eighteenth-century stone statue marks the entrance to one part of the camellia collection.

OPPOSITE: The Huntington mausoleum traces its roots back to classical Greece with its Ionic columns, to classical Rome with its shallow-profile dome like that of the Pantheon, and to Renaissance Italy with its form reminiscent of Bramante's sixteenth-century Tempietto in Rome.

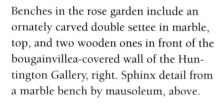

Benches in the rose garden include an ornately carved double settee in marble, top, and two wooden ones in front of the bougainvillea-covered wall of the Huntington Gallery, right. Sphinx detail from a marble bench by mausoleum, above.

Wooden benches are strategically placed for special views throughout the gardens, like the one, top, overlooking a collection of golden barrel cacti in the desert garden and the one at right in the Shakespeare garden. A concrete seat, left, was built in to the hillside in a shady glade in the Japanese garden. This white marble bench, center, is tucked into a garden nook next to the mansion and offers respite to a weary visitor.

Ferocactus latispinus

Det. by: Héctor M. Hernández   Date: April 1992
Collector, No. & Date:

Ferocactus

MEXICO: Puebla. Rte 135, 5.6 km S of
Coxcatlan, along roadside. Alt. 1050m

RESEARCH AND EDUCATION
THE GARDENS' IMPORTANT MISSION

One of the nine thousand specimens in the herbarium, this *Ferocactus latispinus,* a kind of barrel cactus, was collected along a road in the Mexican state of Puebla in 1985.

Red wagons are the conveyances of choice during the popular annual plant sale in May. The Huntington propagates plants for sale all year long. Garden enthusiasts come from long distances to buy plants not always found in nurseries.

MORE THAN A BEAUTIFUL garden-park with interesting planting, the Huntington Botanical Gardens aims to advance the understanding and appreciation of plants and gardens, a mission defined in 1926 by founder Henry E. Huntington.

Specific goals are to maintain the historical character of the gardens, especially those of traditional design around the Library and Huntington Gallery, to interpret estate landscaping and aesthetics as Huntington envisioned them, to preserve some of the original estate grounds, and to present a continuing horticultural display of plants for Southern California gardens. Programs, exhibits, literature, and new collections, sometimes developed with participation from the library and art departments, reinforce the institution's interest in history, literature, and art, while similar efforts address public education in plant biology, the importance of biodiversity, and appreciation and conservation of plants.

The botanical division acquires and curates plant collections, emphasizing those related to existing theme gardens. They test and distribute new plant material, with a special interest in little-known and endangered species, in developing ideas about plant cultivation and landscaping, and in conserving plant biodiversity. They collaborate with

The herbarium specimen, left is *Echeveria difractens.*

The display at right shows how a succulent can be propagated: leaves grow roots and then additional leaves form to make rosettes. This is a *Graptopetalum pentandrum* ssp. *superbum.*

scientists and professionals in plant study and make resources and plant material available to researchers and institutions. Staff members do research, concentrating on areas related to the Huntington's gardens, and publish the results.

Research was an early priority and William Hertrich tested many kinds of plants for suitability to Southern California, including the first successful outdoor planting of orchids. His records of plant survival and losses during major freezes and other data are still useful today.

In 1963 Hertrich's successor Myron Kimnach established a herbarium specializing in plant groups and geographic regions related to the gardens. It provides a permanent record of the gardens' plants, documents collecting expeditions, and holds specimens for taxonomic study. Succulent plants make up its most numerous category; others are plants from the New World tropics, from the world's arid regions, and plants of horticultural importance from wild collections.

Today the gardens attract researchers in botany, taxonomy, and plant physiology, molecular and cellular biology, and hybridization. University and independent scientists have used the plant and herbarium collections for such varied projects as monographs on the genera *Chamaedorea* and *Conophytum*; anatomical research on columnar Mexican cacti, on the wood of Didiereaceae, and on *Welwitschia*; analysis of the latex of euphorbias for industrial applications; and extraction and cloning of *Welwitschia* RNA for enzymes important to cone development.

The garden periodically publishes an *Index Seminum*, a list of seeds native to Southern California and from the gardens' exotic plants, for exchange with other institutions around the world. They in turn receive more than 240 such lists, some of them seed sources for the Huntington.

LEFT: Arms loaded with plant materials, including huge bamboo sheaths, students head for the school bus after a garden class.

RIGHT: Students pick plant specimens like this succulent on docent-led class in the gardens called "Reading Plants," a comparative study of desert, jungle, and aquatic environments.

Educational activities fall into two main categories, interpretation of the gardens and public programs. On-site garden programs reach school children, home gardeners, horticulturists, botanists, and others. For example, a Reading Plants class introduces fourth through seventh graders to differences among plants of the jungle, desert, and aquatic gardens. Another explores the cultural significance of various aspects of the Japanese garden. University instructors may use the gardens as an outdoor classroom or laboratory. And the gardens provide intern work-study experiences for college students and young professionals in horticulture and botany.

Other programs include annual symposia on the study and cultivation of succulents and roses. Lectures and plant sales introduce enthusiasts and home gardeners to landscaping, plant culture and selection, and historical and modern plant varieties. Horticultural societies and garden clubs present public shows of azaleas, camellias, succulents, bonsai, and flower arranging. And a corps of trained docents lead regularly scheduled garden tours, or booklets are available for visitors who prefer to structure their own exploration of this botanical wonderland.

• THE BOTANICAL LIBRARY •
BOOKS RARE AND OLD, NEW AND FAMILIAR

This etching of *Rosa gallica* is one of ninety hand-colored plates in the first book devoted just to roses, *A Collection of Roses from Nature* by Mary Lawrence (1797).

RIGHT: Hand-painted bird of paradise *(Strelitzia reginae)* by Syd Edwards appeared in *The New Flora Britannica*, published in London in 1806.

THE HUNTINGTON'S botanical library consists of a working reference library in the botanical division offices and a collection of rare and important volumes on horticulture and botany kept in the stacks of the Library. In building one of the world's finest libraries in the fields of English and American history, literature, and science, Henry Huntington collected rarities in botanical books mainly when he bought whole collections and private libraries that included them. Since his death, acquisitions of other rare and valuable books have created a sizable special collection in the early history of science, some of it occasionally displayed in the Library's exhibition hall.

The earliest materials are from the fifteenth-century printing revolution that followed the appearance of Gutenberg's Bible, the first book printed with movable type. How botany as a science eventually developed from general knowledge and speculation can be traced in early treatises from first editions of Linnaeus, Darwin, and others. The collection is especially strong in herbals of the sixteenth and seventeenth century, among them works by Otto Brunfels and Leonhart Fuchs, pioneers in drawing from live plants to

MAGNOLIA maximo flore foliis subtus ferrugineis. Rara Hort Osff p. 232.

a. Narcissus polyanthos calice medio luteus. b. Narcissus minor polyanthos, medio luteus. c. Narcissus totus, albus major cum pluribus floribus. d. Narcissus septentrionalis calice luteo pleno duplicato foliis. e. Narcissus septentrionalis albus luteo pleno oris impar. f. Narcissus Orientalis calice aureo pluribus foliolis pleno.

White magnolia is one of 100 color plates from drawings by George D. Ehret in *Plantae Selectae* issued in Germany 1750-73 and published by the botanist, bibliographer, and eminent physician Christoph Trew.

RIGHT: A color illustration of narcissus flowers from *Phytanthosa Iconographia* (1737-45) by Johann Weinmann.

illustrate their books. Such early works on plants and gardens as the first book on the oak tree, by Duchoul, and Jean Robin's floralegium, or compendium of known American and New World plants, also link the Huntington's interest in literature, art, and botanical material. So do eighteenth- and early nineteenth-century botanical garden catalogs that document the emergence of exotic and New World plants and manuals from the same period that instruct the artist in flower drawing and painting. Succulent plants are treated in rare editions of classical works by Engelmann, Haworth, and Lemaire, names perpetuated in plant names. Two seminal editions on roses are the eighteenth-century A *Collection of Roses from Nature*, by Mary Lawrence, the first book devoted exclusively to the rose, and the 1820 *Les Roses*, by Pierre Joseph Redouté, famous for its illustrations.

The library has a complete collection of *Curtis's Botanical Magazine*, now *Kew Magazine*, starting from the late eighteenth century. The first, greatest, and longest continuously published botanical magazine, it chronicles a dazzling array of discoveries from the pre-Victorian era to our time. In the same company are the sumptuously illustrated, nineteenth-century *Edwards' Botanical Register* and Sir Joseph Paxton's *Magazine of Botany*, along with the later *Gardener's Chronicles*.

The botanical reference library, in daily use by the staff, contains important general books and journals but emphasizes materials useful for the identification,

study, and cultivation of plants in the Huntington's major gardens and collections. Areas of concentration include camellias, roses, and succulents; Australia, southern Africa, and Japan; tropics, especially of the New World; and dry-lands floristic and ecological studies. The library also contains gardening, landscaping, culture, botanical, horticultural, and taxonomic general reference materials.

The library of the Cactus and Succulent Society of America is housed here. Newsletters from plant societies across America and "round robin" letters, which contain detailed information on plants for specific climates, propagation, growing tips, and the like, are on file. So are correspondence and ephemera from botanists and seed and nursery catalogs from the nineteenth century on.

CALENDAR OF COLOR *A Changing Pageant of Flower and Foliage Displays Every Season of the Year*

WINTER

The Tournament of Roses, perhaps the world's greatest annual floral spectacle, launches the New Year in nearby Pasadena. In the Huntington's rose garden, pruning is put off until after that so holiday visitors can enjoy the season's last flowers. In January camellias come into their own as reticulatas and then japonicas follow sasanquas already in bloom. Wisterias soon burst forth on arbors in the Japanese garden, while fruit trees flower pink and white there and in the Shakespeare garden. The showiest color of all explodes in the desert garden as aloes produce their red, orange, and yellow flower clusters. Acacias and other Australian trees and shrubs come into flower, while South African bulbs begin year-long color on the ground. Deciduous magnolias begin blooming, as do showy Hong Kong orchid, Taiwan cherry, pink snowball, coral, and other trees.

SPRING

A transition from winter's bright and sunny or sometimes rainy days to the frequently overcast days of late spring, April brings out peak bloom in several gardens. Roses are in full flower, the Shakespeare garden—a color showcase like the entry pavilion's south garden—sports its finest blooms, the North Vista is enlivened with camellias and azaleas, the herb garden has hollyhock and other flowers, and in the Australian garden the curious little kangaroo paws makes its spring appearance. Such flowering trees as bottlebrush, cassias, cape chestnut, and grevillea stand out as individuals while an avenue of jacaranda in the parking lot garden creates a blue-purple color mass. In the desert garden, yuccas and similar plants put forth big flower stalks, cacti show flashy blooms, and beds of flowering ground-cover succulents dazzle with some of nature's most intense hues.

Trumpet tree (*Tabebuia chrysotricha*)

Chinese fringe tree (*Chionanthus retusa*)

Kaffir lily (*Clivia miniata*)

Camellia hybrid, 'Beverly Baylies'

Floribunda rose, 'Orangeade'

Opuntia littoralis

Aloe africana

Azalea

Prunus 'Pink Cloud'

SUMMER

Beds of lavender agapanthus spread color across the ground along with cannas, daylilies, and South African bulbs and other flowers in the subtropical garden. The evergreen Delavay and grandiflora magnolias bloom white, casting heavy scent on the summer air. In front of the Library bright red flowers of Bidwill's coral tree attract hummingbirds by the drove. In other gardens color comes from lacebark bottle tree, cassias, Moreton Bay chestnut, crape myrtle, oleanders, and—toward the end of summer—pink and white silk floss trees. If it seems hot in the desert garden, where yuccas and puyas still bloom in July, seek out the cooler lily ponds, where lotus and water lilies are in full flower, and the jungle garden next to that area, where bright red gingers and hydrangeas bloom in the shade of tall trees.

Bird of paradise (*Strelitzia reginae*)

Daylily (*Hemerocallis*)

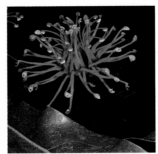

Firewheel tree (*Stenocarpus sinuatus*)

Cereus huntingtonianus

AUTUMN

Heralds of fall, the sasanquas are the first camellias to bloom, from October on. Look for foliage color in the parking lot garden and other spots, but mostly in the Japanese garden, where maples take on varied hues and the Zen garden's grove of ginkgo turns a bright yellow. Other yellows come from flowering cassia trees in the Australian, subtropical, and parking lot gardens. The last roses of summer hang on in their own garden while in December the first aloes and crassulas begin their winter color in the desert garden and the purple Mexican sage puts on its own show in the subtropical garden. Lending yuletide notes, hollies are laden with bright red berries in the Shakespeare and other gardens and both red and white poinsettias flaunt their colorful bracts near the south terrace of the Huntington Gallery.

Gladiolus dalenii

Sweet olive (*Osmanthus fragrans*)

Mexican sage (*Salvia leucantha*)

Airplane plant (*Crassula falcata*)

FURTHER READING

ABOUT THE HUNTINGTON

An Introduction to the Huntington Botanical Gardens (pamphlet). San Marino: Huntington Library.

Bernal, Peggy Park. *The Huntington. Library, Art Collections and Botanical Gardens.* San Marino: Huntington Library, 1992.

Dickinson, Donald C. *Henry E. Huntington's Library of Libraries.* San Marino: Huntington Library, 1995.

The Founding of the Henry E. Huntington Library and Art Gallery: Four Essays. San Marino: Huntington Library, 1977.

Garden Notes. The Huntington Botanical Gardens. San Marino: Huntington Library, 1969.

Hertrich, William. *The Huntington Botanical Gardens 1905-1949. Personal Recollections of William Hertrich, Curator Emeritus.* San Marino: Huntington Library, 1949.

Schad, Robert O. *Henry Edwards Huntington. The Founder and the Library* (pamphlet). San Marino: Henry E. Huntington Library and Art Gallery, 1987.

Spurgeon, Selena A. *Henry Edwards Huntington. His Life and His Collections.* San Marino: Huntington Library, 1992.

The Huntington Art Collections: A Handbook. San Marino: Huntington Library, 1986.

Thorpe, James. *Henry Edwards Huntington: A Biography.* Berkeley: University of California Press, 1994.

—————. *Henry Edwards Huntington, A Brief Biography* (pamphlet). San Marino: Huntington Library, 1995.

HISTORY

Dumke, Glenn S. *The Boom of the Eighties.* San Marino: Huntington Library, 1944.

Farnsworth, R. W. C. *A Southern California Paradise.* Pasadena, 1883.

Garnett, Porter. *Stately Homes of California.* Boston: Little, Brown, and Company, 1915.

Holder, Charles F. *All About Pasadena and Its Vicinity.* Boston: Lee & Shepard, 1889.

James, George Wharton. *Tourists' Guide Book to South California.* Los Angeles: B. R. Baumgardt & Co., 1895.

Johnson, Paul C. *The Early Sunset Magazine 1898–1928.* San Francisco: California Historical Society, 1973.

Lyon, William S. *Gardening in California. A Brief Treatise on the Best Methods of Cultivating Common Flowers in the California Home Garden.* Los Angeles, 1904.

Padilla, Victoria. *Southern California Gardens.* Santa Barbara: Allen A. Knoll, Publishers, 1994.

Sanborn, Kate. *A Truthful Woman in Southern California.* New York: D. Appleton & Company, 1893.

Saunders, Charles Francis. *The Story of Carmelita.* Pasadena: A. C. Vroman, Inc., 1928.

Sherwood, Mildred. *Days of Vintage, Years of Vision.* 2 vols. San Marino: Orizaba Publications, 1982-1987.

Truman, Major Ben C. *Semi-Tropical California.* San Francisco: A.L. Bancroft and Company, 1874.

Writers' Program. *Los Angeles. A Guide to the City and Its Environs.* Comp. by Workers of the Writers' Program. New York: Hastings House, 1951.

PERIODICALS

BCI Magazine. Tallahassee: Bonsai Clubs International.

Sunset Magazine. San Francisco, 1900–1912.

GENERAL HORTICULTURE

There are many good books about gardening; these are a few favorites of the Huntington's botanical staff.

Bailey, L. H. *Hortus Third.* New York: Macmillan, 1976.

Everett, Thomas H. *The New York Botanical Garden Illustrated Encyclopedia of Horticulture.* 10 vols. New York: Garland, 1980.

Harris, Richard W. *Arboriculture.* Englewood Cliffs, NJ: Prentice-Hall, Inc., 1983.

Mabberley, D.J. *The Plant-Book.* New York: Cambridge University Press, 1993.

Mathias, Mildred E., ed. *Flowering Plants in the Landscape.* Berkeley: University of California Press, 1982.

Menninger, Edwin A. *Flowering Vines of the World.* New York: Heathside, 1970.

The New Royal Horticultural Society Dictionary of Gardening. New York: Stockton Press, 1992.

Polunin, Oleg. *Flowers of Europe.* London: Oxford University Press, 1969.

Sunset Western Garden Book. Menlo Park: Lane Publishing Co., 1995.

Tustin, Thomas Gaskell, ed. *Flora Europaea.* 5 vols. Cambridge: Cambridge University Press, 1964.

AUSTRALIAN PLANTS

Brooker, M. I. H. and R. C. Barneby. *Field Guide to Eucalyptus.* 2 vols. Melbourne: Inkata Press, 1983.

Elliot, W. R., and D. L. Jones. *Encyclopedia of Australian Plants Suitable for Cultivation.* 6 vol. Melbourne: Lothian, 1980-1993.

Flora of Australia. Vol. 19: Eucalyptus, Angophora. Canberra: Aust. Govt. Pub. Service, 1988.

Kelly, Stan. *Eucalypts.* 2 vols., 2nd ed. New York: Van Nostrand Reinhold Co., 1983.

Wrigley, John W., and Murray Fagg. *Australian Native Plants.* 3rd. ed. Sydney: Collins, 1988.

BAMBOO

Dajun, Wang, and Shen Shap-jin. *Bamboos of China.* London: Christopher Helm, 1987.

Suzuki, Sadao. *Index to Japanese Bambusaceae.* Tokyo: Gakken Co., Ltd., 1979.

CACTI AND OTHER SUCCULENTS

Benson, Lyman. *The Cacti of the United States and Canada.* Palo Alto: Stanford University Press, 1982.

Cullman, Willy, et al. *The Encyclopedia of Cacti*. Dorset, Sherborne, England: Alphabooks, 1986.

Folsom, Debra Brown, and John Trager. *Dry Climate Gardening with Succulents*. New York: Pantheon, 1995.

Gentry, Howard S. *Agaves of Continental North America*. Tucson: The University of Arizona Press, 1982.

Hertrich, William. *A Guide to the Desert Plant Collection*. San Marino: Huntington Botanical Gardens, 1937.

Jacobsen, Hermann. *The Lexicon of Succulent Plants*. London: Blanford Press, 1970.

Lyons, Gary W. *The Huntington Desert Garden*. Reseda: Abbey Garden Press, 1975.

Pilbeam, John. *Cacti for the Connoisseur*. Portland: Timber Press, 1987.

Rowley, Gordon. *The Illustrated Encyclopedia of Succulents*. New York: Salamander Book, Crown Publisher, 1978.

Reynolds, G.W. *The Aloe of South Africa*. Rotterdam: A.A. Balkema, 1982.

———. *The Aloes of Tropical Africa and Madagascar*. Mbabane, Swaziland: Aloes Book Fund, 1986.

Self-Guided Tour to the Huntington Desert Garden (pamphlet). San Marino: Huntington Library.

CAMELLIAS

Bliss, Amelia, and Carey S. Bliss. *Camellias. The Huntington Gardens* (pamphlet). San Marino: Huntington Library.

Feathers, David L., ed. *The Camellia*. Columbia, SC: R. L. Bryan Co., 1978.

Hertrich, William. *Camellias in the Huntington Gardens*. 3 vols. San Marino: Huntington Botanical Gardens, 1954.

CYCADS

Hertrich, William. *Palms and Cycads, As Observed Chiefly in the Huntington Botanical Gardens—Their Culture in Southern California*. San Marino: Henry E. Huntington Library and Art Gallery, 1951.

Jones, David L. *Cycads of the World*. Washington, D.C.: Smithsonian, 1993.

Stewart, Lynette. *A Guide to Palms & Cycads of the World*. Sydney: Angus & Robertson, 1994.

HERBS

MacGregor, John C., IV. *A Guide to the Huntington Herb Garden*. San Marino: Huntington Library, 1983.

JAPANESE GARDENS

Makino, Tomitaro. *Makino's New Illustrated Flora of Japan*. Tokyo: The Hokuryukan Co., 1961.

Naka, John. *Bonsai Techniques*. 2 vols. Santa Monica: Dennis-Landman for Bonsai Institute of California, 1982.

Ohwi, Jisaburo. *Flora of Japan*. Washington, D.C.: Smithsonian Institute, 1965.

Self-Guided Tour to the Huntington Japanese Garden (pamphlet). San Marino: Huntington Library, 1985.

JUNGLE PLANTS

Jones, David L. *Encyclopedia of Ferns*. Portland: Timber Press, 1987.

Rauh, Werner. *Bromeliads for Home, Garden & Greenhouse*. Poole, England: Blandford Press, 1979.

PALMS

Hertrich, William. *Palms and Cycads, As Observed Chiefly in the Huntington Botanical Gardens—Their Culture in Southern California*. San Marino: Henry E. Huntington Library and Art Gallery, 1951.

Jones, David. *Palms in Australia*. Frenchs Forest, NSW: Reed Books, 1984.

Stewart, Lynette. *A Guide to Palms & Cycads of the World*. Sydney: Angus & Robertson, 1994.

Uhl, Natalie W., and John Dransfield. *Genera Palmarum*. Lawrence, Kansas: Allen Press, 1987.

ROSES

Austin, David. *Old Roses and English Roses*. Woodbridge, Suffolk: Antique Collectors Club, 1990.

Cairns, Thomas, ed. *Modern Roses 10*. Shreveport, LA: American Rose Society, 1993.

Martin, Clair. *English Roses in Southern California*. Pasadena: Hortus Gardenbooks, 1993.

Phillips, Roger, and Martyn Rix. *Roses*. New York: Random House, 1988.

Thomas, Graham Stuart. *The Graham Stuart Thomas Rose Book*. Portland: Timber Press, 1994.

SUBTROPICAL PLANTS

Bramwell, David, and Zoe I. Bramwell. *Wild Flowers of the Canary Islands*. London: Stanley Thornes Ltd., 1974.

Bryan, John E. *Bulbs*. 2 vols. Portland: Timber Press, 1989.

Du Plessis, Niel, and Graham Duncan. *Bulbous Plants of Southern Africa*. Cape Town: Tafelberg, 1989.

Elivoson, Sima. *Wild Flowers of Southern Africa*. Johannesburg: Macmillan, 1980.

Palgrave, Keith Coates. *Trees of Southern Africa*. Cape Town: C. Struick, 1984.

Polunin, Oleg, and Anthony Huxley. *Flowers of the Mediterranean*. London: Chatto & Windus, 1965.

Rix, Martyn, and Roger Philips. *The Bulb Book*. London: Pan Books Ltd, 1981.

Vogts, Marie. *South Africa's Proteaceae*. Cape Town: C. Struik, 1982.

TREES

Callaway, Dorothy. *World of Magnolias*. Portland: Timber Press, 1994.

Gelderen, D. M. van, and J. R. P. van Hoey Smith. *Conifers*. 2nd ed. Portland: Timber Press, 1989.

Hodel, Donald R. *Exceptional Trees of Los Angeles*. Chatsworth, CA: California Arboretum Foundation, 1988.

Krüssmann, Gerd. *Manual of Cultivated Broad-leaved Trees & Shrubs*. 3 vols. Beaverton: Timber Press, 1984.

——. *Manual of Cultivated Conifers*. Portland: Timber Press, 1985.

Menninger, Edwin A. *Flowering Trees of the World*. New York: Heathside, 1962.

Stubley, Wendy. *Trees of San Marino*. San Marino: Henry E. Huntington Library and Art Gallery, 1989.

Treseder, Neil G. *Magnolias*. London and Boston: Faber and Faber, 1978.

INDEX

Page numbers in bold refer to illustrations

THE BOTANICAL GARDENS AT THE HUNTINGTON
was produced by Perpetua Press, Los Angeles
Designer: Dana Levy
Editor: Letitia Burns O'Connor
Indexer: Kathy Talley Jones
Typeset in Berkeley Oldstyle and Michelangelo
Printed in Singapore
by Toppan Printing Company